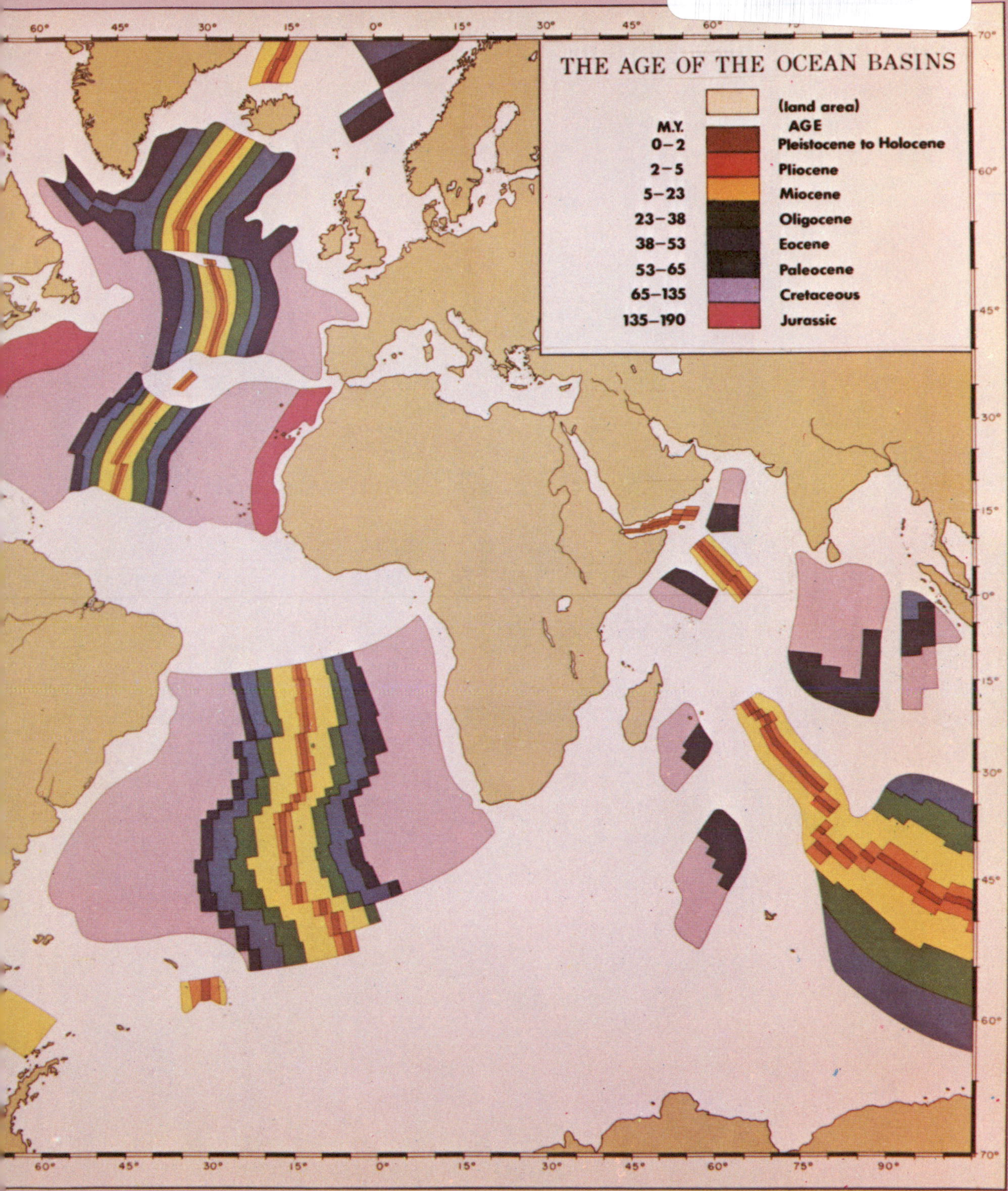
THE AGE OF THE OCEAN BASINS
(land area)
M.Y.
AGE
0–2 Pleistocene to Holocene
2–5 Pliocene
5–23 Miocene
23–38 Oligocene
38–53 Eocene
53–65 Paleocene
65–135 Cretaceous
135–190 Jurassic
60° 45° 30° 15° 0° 15° 30° 45° 60° 75° 90°
70° 60° 45° 30° 15° 0° 15° 30° 45° 60° 70°

Color map, *The Age of the Ocean Basins,* compiled by Walter C. Pitman III, Roger I. Larson, and Ellen M. Herron of Lamont-Doherty Geological Observatory. Copyright 1974, the Geological Society of America, Inc. This map attempts to match the ages of the ocean basin on the two sides of the mid-ocean range. Ages are given in millions of years.

Evolution of Coasts, Continental Margins, & the Deep-Sea Floor

Crane, Russak & Company, Inc.
New York

Geological Oceanography

Francis P. Shepard

Geological Oceanography:
Evolution of Coasts, Continental Margins, and the Deep-Sea Floor

Crane, Russak & Company, Inc.
347 Madison Avenue
New York, New York 10017
ISBN 0-8448-1064-9
LC 76-54533

Printed in the United States of America

Contents

Geological Time Scale

Era	Period	Epoch	Age in years
Cenozoic	Quaternary	Recent	10,000
		Pleistocene	1,800,000
		Pliocene	5,000,000
		Miocene	22,500,000
	Tertiary	Oligocene	37,500,000
		Eocene	53,500,000
		Paleocene	65,000,000
Mesozoic	Cretaceous		140,000,000
	Jurassic		200,000,000
	Triassic		230,000,000
Paleozoic	Permian		280,000,000
		Pennsylvanian	
	Carboniferous	Mississippian	345,000,000
	Devonian		390,000,000
	Silurian		430,000,000
	Ordovician		500,000,000
	Cambrian		570,000,000
Pre-Cambrian			

Preface

Eighteenth-century geologists found that most of the sedimentary rocks on the continents had been deposited in the oceans of the past. But it was only towards the middle of the twentieth century that geologists began to study the geology of the sea on a sufficiently large scale to find out what has been going on in our present day oceans, both to explain the origin of marine sedimentary rocks and to make a serious attempt to trace the history of the oceans during the long geological past. Ever since, as the great void existing in the science of geology was realized, there has followed a flood of new literature pouring into the scientific journals, growing on an exponential curve until it is almost hopeless to try to keep abreast of new facts and ideas being brought forth. This book is an attempt to give to students and the interested public the gist of what we have been finding in this new field. Throughout the book non-technical terms and simplified explanations are used to explain the findings of the highly technical books and journals.

Included are the revolutionary new ideas that have come from the marine investigations by geologists and geophysicists who have been covering all of the oceans of the globe using new devices and techniques that were not even dreamed of 50 years ago. Not only has this new flood of information vastly altered the ideas scientists have concerning the history of the earth, but also it appears to have an enormous effect on the world economy and on our industrial potential for the future. Resources that were thought to be nearing exhaustion in our time are being found in great quantities on and below the ocean floor.

Throughout this book, distances are given in feet and statute miles, but, in order to help the conversion to the metric system, distances are also given in parenthesis in meters and kilometers. Some depths are given in fathoms, which are equal to six feet, and these have been converted to meters. However, in some places it was appropriate to use only nautical miles, one mile equalling 6,080 feet.

Helpful suggestions were made by my colleagues Sir Edward Bullard (for Chapter 2), Dr. Robert Arthur (for Chapter 3), and Dr. E. L. Winterer (for Chapter 11). My wife, Elizabeth Shepard, was helpful with the style of the manuscript, and Patrick McLoughlin and Neil Marshall made good suggestions for the illustrations. Nance North was particularly helpful with the bibliography.

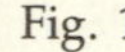

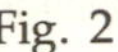

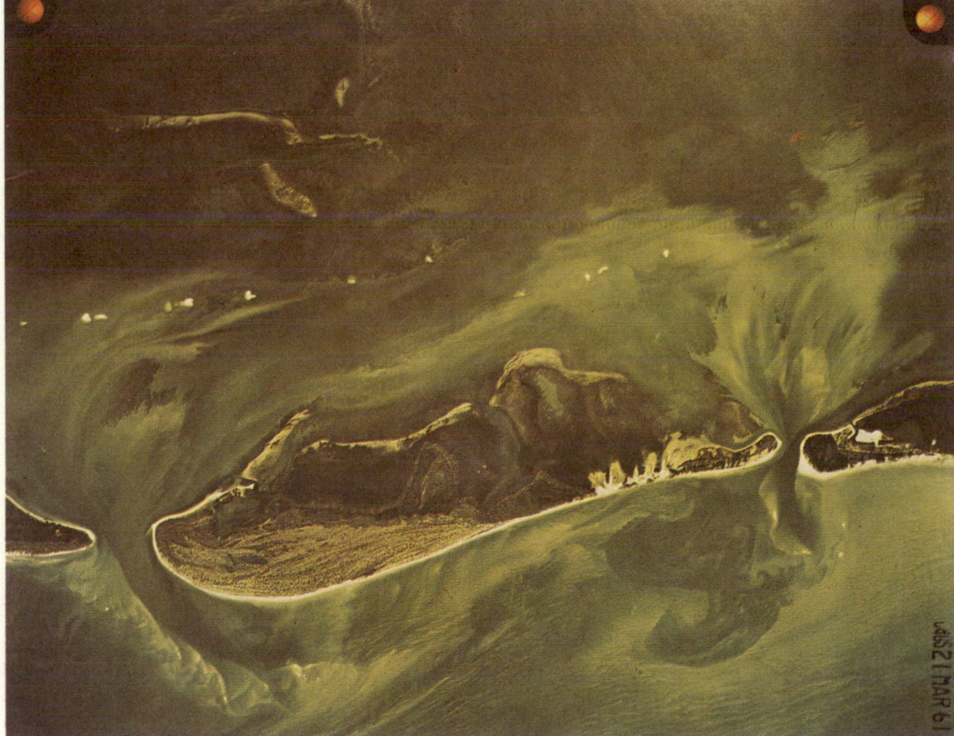

Plate 1

Fig. 1. Barrier island off Nauset Beach, Cape Cod. The barrier was cut through by the hurricane of 1961 and a fan was deposited into the lagoon. Subsequently, a new beach was built on the outside. Photo by U.S. Coast and Geodetic Survey, 1962.

Fig. 2. Showing the deep tidal channels between barrier islands along the southwest coast of Florida. Sanibel Island in center. Photo by U.S. Coast and Geodetic Survey, 1961.

Fig. 3

Fig. 4

Plate 1

Fig. 3. Bora Bora, Society Islands, showing the partly eroded central volcano and a volcanic coast bordered by a lagoon and a barrier coral reef.

Fig. 4. Remnants of wave erosion at Cape San Martin on the California coast. The stacks are covered with guano, which is a valuable resource actually mined from the small islands off the coast of Peru.

Fig. 1

Fig. 2

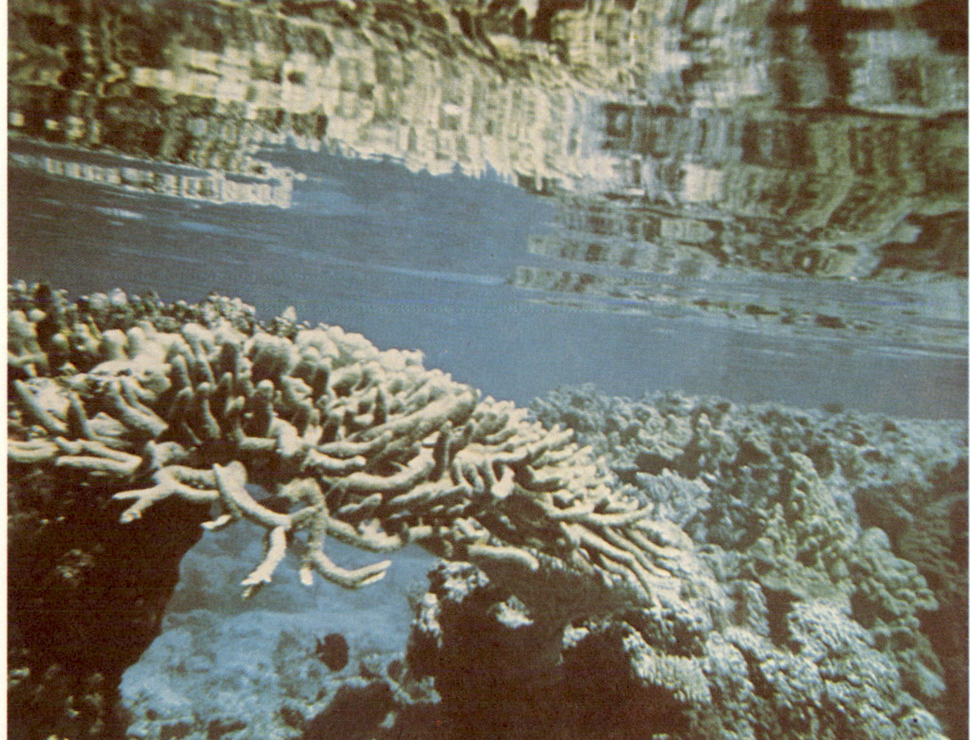

Plate 2

Fig. 1. Actively growing coral reef on the inner edge of the barrier reef off western Moorea, Society Islands.

Fig. 2. A growing reef off the west coast of Moorea with a frond of *Acrapora* forming an arch between two high-standing coral clusters. The upper portion of the photo is a reflection of the reef in the underwater surface.

Fig. 3

Fig. 4

Plate 2

Fig. 3. A partly dead coral reef along the coast of Bora Bora. Note the *Tridacna* clam (blue), which had bored its way into the coral reef.

Fig. 4. The large *Acrapora palmata* coral fronds that are so abundant in the West Indies and Bahama reefs. These are rather easily broken off by large storm waves. Photographed off southeast Martinique Island.

Introduction

Chapter 1

Just before getting his second prize from the Geological Society of America, Nobel prize-winner Harold C. Urey recalled that when he studied geology in the early part of the century he really did not like it. One is led to wonder how a man who has had such an impact on science and covered such a broad field of interest could have failed to have been aroused by his early contact with geology, which involves so much exciting history of the development of the earth into its present state. But geology was not such a dynamic subject in those days as now. Geology had reached a stage of development when a few famous professors were convinced that they had solved the important problems of the science, and were teaching their students the law and gospel as if it were a religion. I remember in my student days feeling depressed that I had not become a geologist when there were still many problems to be solved. This, of course, was before geology underwent its great revolution. Quite suddenly during the 1960s, we found ourselves discarding most of our philosophy of the orderly development of the planet, and taking up what at first seemed a prophetic dream of continents splitting apart and new oceans forming. Suddenly many of the puzzles of geological history began to make sense, and we began to discard most of the ideas that we had thought were so well established. Today, with these new developments, geology has really become an exciting subject.

The new concepts have developed to a great extent because geologists working for centuries on land finally decided to go to sea and find out what was under the unexplored 72 percent of the earth's surface. Before World War I, marine geology was almost unknown. In 1923, when I first began on a very small scale to study the sea floor, I was literally the only geologist in the United States, perhaps in the world, who was working in this field. The old textbooks generally had a chapter about the ocean, but a large part of it was based on speculation that dealt little with facts. One reads in these old texts that most of the ocean floor is montonously flat, like the great plains of the central United States. When soundings were hundreds of miles apart, as they were before the days of echo soundings, there was no way to question this concept; but in 1920 echo soundings were invented, and this allowed us to make continuous profiles of the sea floor without even stopping the ship. The ocean floor proved to be as irregular as the land. Similarly, it seemed logical to believe that ocean sediments should be coarse grained near the shore, where the fine sediments are removed by the breaking waves, and fine grained farther from the land. This is how they were described in the old textbooks. When I began collecting sediments on the continental shelf off New England, I was amazed to find that here the sediments, instead of being finer as I got farther from shore, often were coarser. As more and more samples were taken by various investigators, it became evident that this finding off New England was by no means unusual. It began to look as though the ocean floor was more complicated than had been generally believed.

Many other discoveries soon followed. Maurice Ewing, of Lamont-Doherty Geological Observatory, began his innovative career in the 1930s, developing such things as underwater photography using cameras lowered to the bottom. From these photos we began to find evidence that the sea bottom processes were much more active in deep water than had been supposed. Ewing also was a leader in developing sound-transmission methods that gave us our first ideas of what was under the bottom. Some other geophysical methods were used to investigate the ocean floors. Vening Meinesz of Holland started measuring gravity from a submarine. Sir Edward Bullard, of Cambridge University, and Roger Revelle, of Scripps Institution of Oceanography, began taking heat measurements from probes stuck into the bottom, and Victor Vacquier, now of Scripps Institution, started measuring the magnetism of the sea floor. All of these methods gave startling results.

Finally, scientists decided it was time to have a look at the bottom from deep-diving vehicles. In 1945, Auguste Piccard built the Bathyscaphe, essentially an oil-filled sea-going balloon with a diving bell attached, and made the first deep dives to the ocean floor. Jacques Cousteau was soon to follow with his remarkable explorations in the *Diving Saucer.* Other deep-diving vehicles were developed, and at last geologists were given the opportunity to look at the interesting features their instruments had been detecting from surface operations. During World War II, the first development by Cousteau and Emile Gagnan of what is now called scuba diving has allowed many geologists to observe the shallow sea floor for themselves.

Ocean-floor drilling has given us a vast fund of information. We have uncovered old sunken land masses, discovered numerous reversals of earth magnetism, and filled in many gaps in the history of the earth. Unique fossils have been discovered. Areas now in the Northern Hemisphere have been found to have been formerly in the Southern Hemisphere. The drillings have demonstrated that the crust under the oceans has been sliding so actively under the continents that in no place can we drill down into sediments as old as the earliest marine sediments discovered on the continents.

The British *Challenger* Expedition, 1872-76, gave scientists the first indications of the nature of the ocean basins, but little progress was made until World War I aroused interest in sending sound waves through the sea in order to detect the presence of submarines. We now know that sea mammals had been using a similar method for countless ages so they could penetrate murky water and find hidden food. After the war was over, scientists used the echo method to tell the depths of water under ship keels, and echo sounding soon followed. Thus a sounding could be obtained by measuring the time between sending a signal and receiving an echo, which takes a few seconds in the deepest water as compared to half a day by the old vertical cast method of lowering a weight attached to a line and observing on a counter the amount of wire suspended below the ship. Although World War II threatened to wipe out civilization, this war led to many important scientific developments. In testing different sound frequencies to help locate submarines, scientists found that low frequencies were capable of sending sound waves through the sea floor and getting reflections from deeply buried layers of rock. These sound-transmission methods have revolutionized the field of marine geology. Not only has science benefited, but industry, particularly in the search for new energy sources, has been given a new method of locating petroleum. Many localities where adjoining land areas have no sign of oil now have prosperous oil fields on the adjacent sea floor, discovered largely as a result of surveys with underwater sound.

The juncture between the land and the sea, called the shoreline or the coast, is worthy of much study by man, largely because it is subject to so much change. People like to live closer to the ocean, partly because the ocean is a great source of food, and partly because of the beauty of the coastal scenery and the joy of swimming in the sea. Also, the climate is almost always pleasanter along the coast, warmer in winter and cooler in summer. However, much of the habitation along the coast is on the sandy platforms that adjoin the shore, and these are just the places where the shorelines are subject to such rapid change. A great storm or a tidal wave (tsunami) may cause the loss of houses that have been built near the ocean so that the residents could have a fine view of the sea. We are gradually learning to avoid building on these threatened areas, and we are finding ways to provide protection for communities already constructed there. A few years ago we knew very little about the great sea waves that accompanied earthquakes and the great surges of sea onto the land during hurricanes, but now they are well documented and we are learning how to reduce their destruction.

Geology professors used to teach their students that the deep-ocean floor was covered with the limey or siliceous remains of the small animals of the sea, except at great depths where these organic remains had been dissolved and where a reddish clay covered the floor. Now that we have made extensive collections of sediments from the deep oceans we have found that sands are not infrequent among the sediments of the deep-sea floors. As for red clay, it is actually very rare, although a brown clay is a common type of deep-sea sediment. One of the great surprises has been the discovery of extensive zones where the deep seas are paved with rounded nodules of manganese. These contain such a quantity of cobalt and nickel that there are now plans to quarry them from surface ships and giant dredges. It is even thought that this type of mining may largely replace mining on land, at least for these two metals.

In these days of extensive scuba diving and of the use of face masks by swimmers, we are learning much more about the coral reefs that are so abundant in the tropical areas. It is only since World War II that we have fully appreciated the significance of a theory developed by Charles Darwin more than 100 years ago. Darwin conceived the idea that the ringlike coral reefs, known as atolls, are the result of coral growing around ocean volcanoes that were sinking while the reefs grew upward until the volcano peaks had disappeared, leaving only the surrounding reefs. We find that there are many flat-topped mountains in the ocean that are as much as a mile below the surface. These are thought to be sunken shoals that did not have upgrowing coral masses to keep building the platforms to the surface as the shoals sank.

Another exciting discovery of recent years is the numerous oval-topped hills that stand above the deep-ocean floor along the continental margins, particularly on both sides of the Atlantic. They are also located in portions of the deep inland seas that are largely landlocked, notably the Mediterranean and the Gulf of Mexico. As a result of drilling, we have now established that many of these are domes of salt that have been pushed up through the sediment covering the ocean floor, and are comparable to the salt domes that have proven of such enormous value in the search for oil on the continents and the continental shelves. To have salt in the large quantities necessary for the formation of salt domes that are thousands of feet thick and cover large territories means that huge salt lakes or seas existed in a climate that was arid enough to evaporate the water and concentrate its relatively small salt content until it became deposited on the bottom in thick layers. How could this have happened in what was the ancestor of the great oceans? There are many puzzling things about the ocean floor, and it is only recently that we think we have begun to solve many of these mysteries. Clearly, we are in the midst of a great detective story and our new methods in oceanography are bringing in information so fast that we can now speculate on many strange features that formerly seemed almost impossible to comprehend.

Suggested Supplementary Reading

Davis, R. A., Jr., *Principles of Oceanography* (Menlo Park, California: Addison-Wesley Publ. Co., 1972), Ch. 1.

Ross, D. A., *Introduction to Oceanography.* (New York: Appleton-Century-Crofts, 1970), pp. 29-40.

AZORES
CANARY ISLANDS

Forming new ocean basins and sliding of continents

Chapter 2

It is appropriate to start by plunging into the maelstrom that seems to have engulfed most of the geological community in its swirling currents. At first, the hypothesis was called *drifting continents.* There is some uncertainty as to who started the idea, but it was developed chiefly by Alfred Wegener (1924), an Austrian meteorologist whose book *The Origin of the Continents and Oceans* was first published in German in 1912. He maintained that an ancient continent called Pangaea (Fig. 2-1) included virtually all the land area of the earth in the Paleozoic and early Mesozoic eras a few million years ago (see Geology Timetable inside cover). That vast continent was broken apart about 135 million years ago and the fragments constituting North and South America drifted westward from Africa and Eurasia, leaving the Atlantic Ocean in the gap. At about the same time, Antarctica was separated from the northern continents, and the latter drifted northward; a new portion now constituting India split from Africa on the west and Australia on the east. India had the largest drift, finally impinging against a part of southern Asia. Australia drifted to a point south of the east side of the Asiatic continent.

Wegener had many appealing arguments for his hypothesis. The two sides of the Atlantic certainly appear to fit together, as one can see by making a tracing of the East Coast shelf edge of the Americas on a globe and moving the tracing eastward until it reaches the West Coast shelves of Europe and Africa (Fig. 2-2). Furthermore, geologists had known for many years that large-scale glaciation occurred during the Permo-Carboniferous

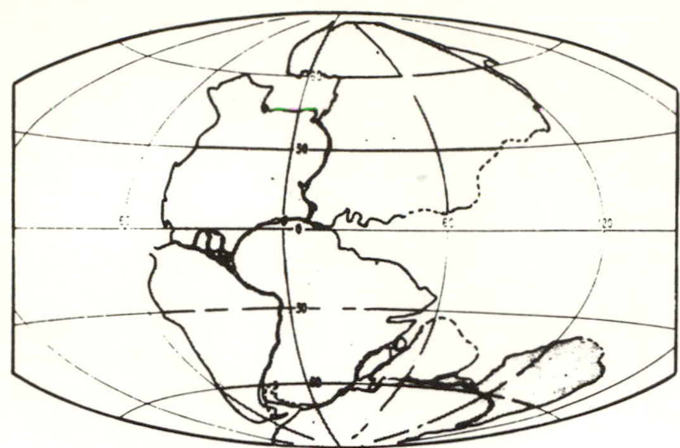

Permian: 255 million years ago

Triassic: 215 million years ago

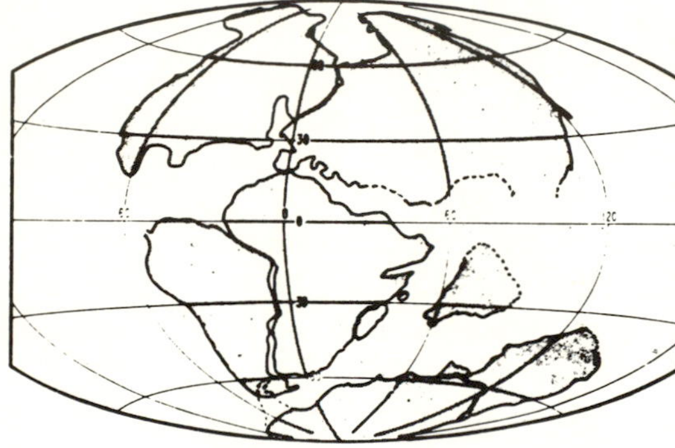

Jurassic: 170 million years ago

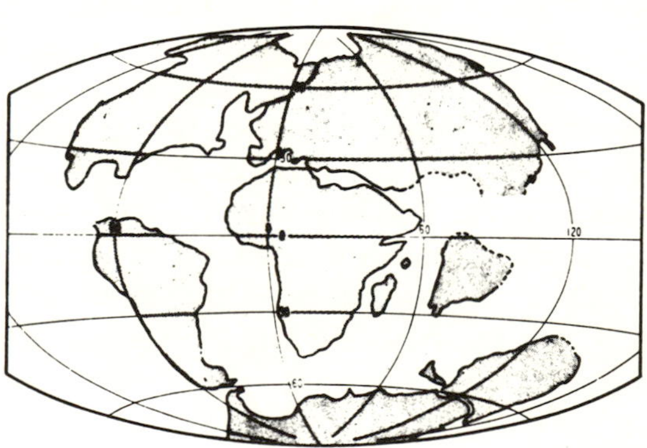

Cretaceous: 100 million years ago

Cenozoic: Present

Fig. 2-1. Showing the breakup of the ancient continent of Pangaea. From R. S. Dietz and J. C. Holden, *Geological Society of America Bulletin*, 1970.

period (about 180 million years ago) in South America, South Africa, India, and Australia. These glaciated lands are now all in low latitudes, most of them in the tropics. Therefore, Wegener concluded that these lands were once

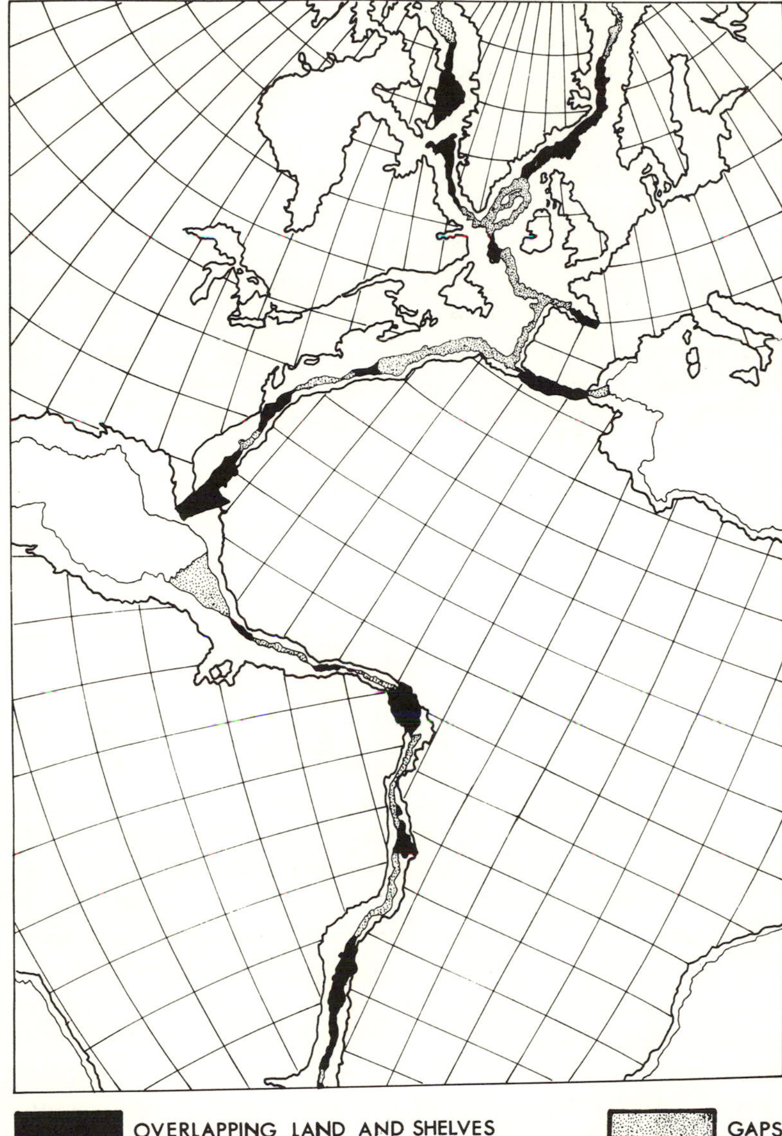

Fig. 2-2. The fit between the two sides of the Atlantic made by tracing the East Coast shelf edge of the Americas on a globe and fitting it to the West Coast shelf edges of Europe and Africa. From Bullard, "The Ocean," *Scientific American*, pp. 66-75, v. 221, no. 3, 1969.

clustered around the South Pole. He also called attention to the great similarity of the fossils of Paleozoic and early Mesozoic ages on the two sides of the Atlantic. Similar matching of fossils of these same ages can be made between South Africa, India, and Australia. Wegener even noted that some of the mountain ranges on the two sides of the Atlantic came to abrupt ends at the continental margins, but the ends could be fitted together at the former supposed continental connections.

The Wegener hypothesis became popular in South Africa and South America but aroused horror among most Americans geologists and many Europeans. How could a great continent move like a ship over the solid earth's interior, particularly for thousands of miles? The fit of the two sides of the Atlantic was denied or called coincidence by some, and by others called a remnant of the early molten stage of the earth when the continents may have consolidated and were moving through the still molten ocean basins like ships in a sea. Fossil similarities on the two sides were explained either by migration across the northern areas where only narrow seas separate the continuity, or by land bridges thought to have later sunk into the deep ocean.

After petroleum geologists had made extensive investigations of South America and Africa and other geologists had begun exploration of Antarctica, many new converts were added to the drifting continent idea, but most geologists held back because of serious unanswered questions. Finally, the impasse was broken in the 1960s, largely as a result of our increasing knowledge of the oceans. Two seagoing American geologists, Harry Hess (1962) of Princeton University and Robert Dietz (1961) of the National Oceanic and Atmospheric Administration, at about the same time suggested a new hypothesis, which Dietz called *Sea-Floor Spreading*. The new discovery that the great ridge running down the center of the Atlantic Ocean has along much of its length a deep, steep-walled valley suggesting the rift valleys of Africa. This indicated that a great crack exists in the middle of the Atlantic Ridge and is being pulled apart, and that lava is flowing into the crack from the molten layer under the ocean crust. Thus the crack becomes temporarily filled and consolidates, resulting in a widening of the ocean floor (Fig. 2-3). This theory has some real differences from the Wegener hypothesis and fits better with new scientific discoveries. Put into its simplest terms, the lava rising into the great continental crack is heavier than the rocks underlying the old continent on either side. Therefore, because of gravity, the lava would come to rest at a lower level than the continents. As a result, after the lava had consolidated, the sea found its way into the depressed portion, forming an elongate narrow body of water, like the Gulf of California or the Red Sea, both now considered as rift valleys. According to the hypothesis, the giant strait with its underlying lava floor was subjected to the same rifting action as the old continent, and a new crack gradually opened in the center of the strait, allowing more lava to flow in as the severed continents were pulled farther apart. Continued cracking and new injections of lava gradually produced a wide ocean, with the original sides of the crack constituting the new continental margins. As the ocean floor spreads, the adjacent continents of each side move with the ocean crust, as in Fig. 2-3. The combined plate of ocean and

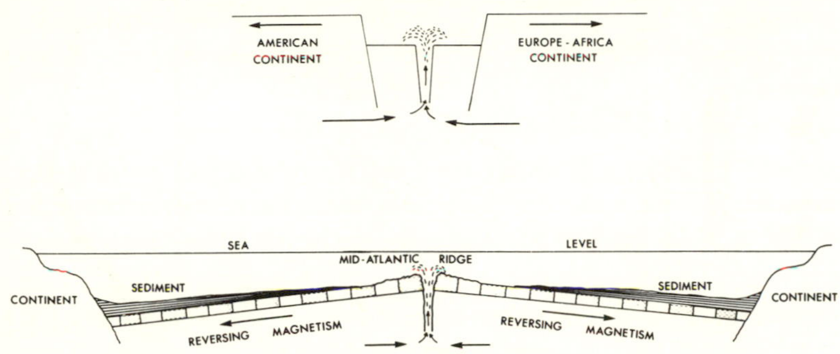

Fig. 2-3. Two stages in the development of the Atlantic according to the sea floor spreading hypothesis. The alternating bands represent the changes between the poles developing positive and negative magnetism in the new ocean crust. Note thickening of the sediments away from the ridge.

continent is thus drifting away from the oceanic ridge, in contrast to the Wegener idea of drifting continents.

How, then, can we account for the Mid-Atlantic Ridge? The molten lavas that come up into the crack are much more expanded than the now cooled older lavas that came up earlier. Therefore, the old lava floor must have sunk. Hence, the new injected oceanic crust would at first lie at rather shallow depths, like the Red Sea, and as the intrusions were pushed aside and were cooling the older lavas would sink to a lower level, leaving the ridge in the center. The ridge, in turn, has elongate valleys in its center except where the cracks have been entirely filled by the intruding lava.

The next problem that the new hypothesis solved was what happened on the other side of the migrating American continents. If we think of this western side as the prow of a moving ship, the obvious answer is that the

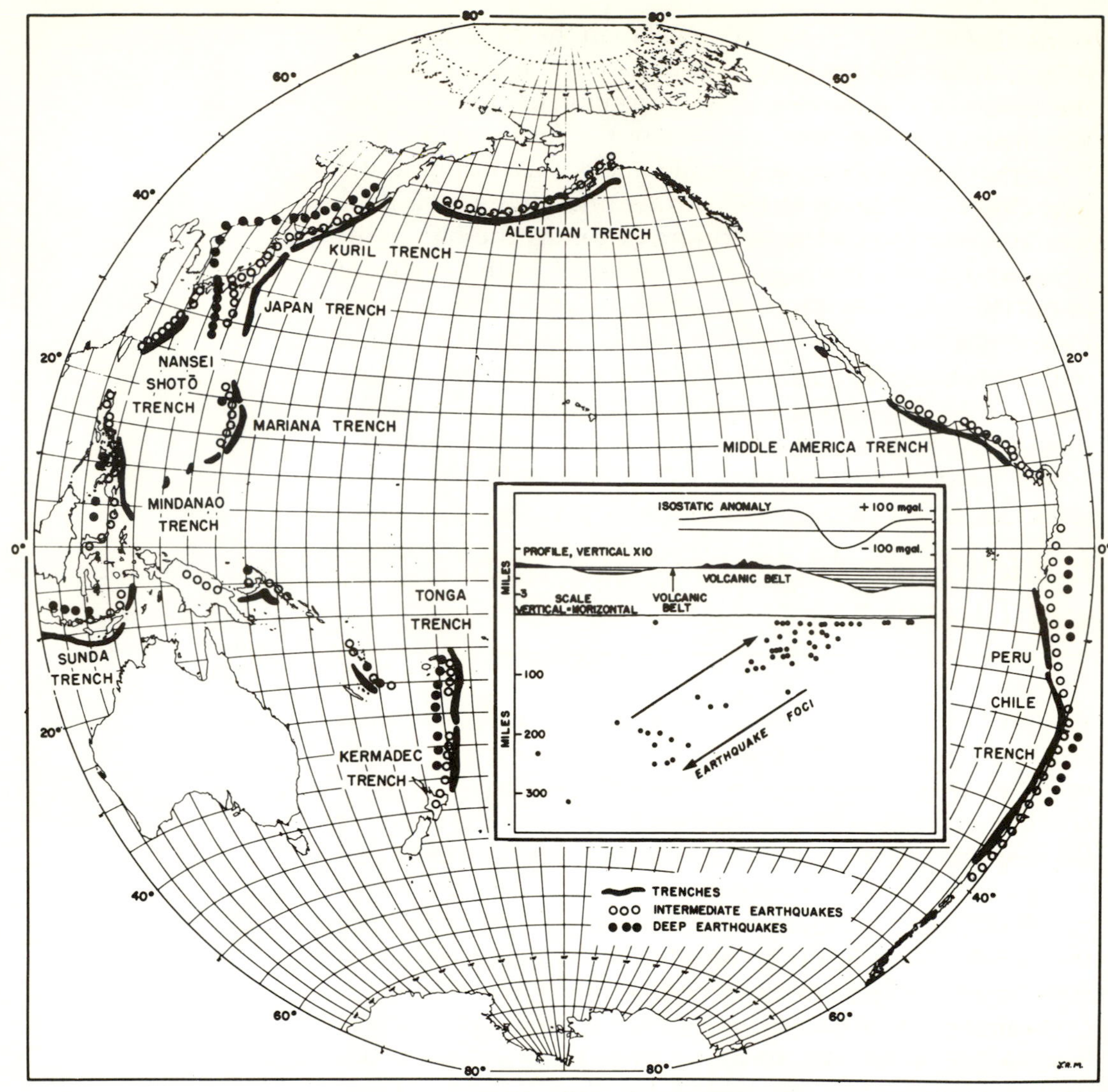

Fig. 2-4. The deep trenches of the Pacific with their shallow-depth earthquakes and the intermediate and deep earthquakes under the adjacent land areas. This is thought to indicate that the ocean crust is being subducted under the continents, as shown by arrows.

crust underneath the ocean into which the continent was advancing must have sunk below the migrating continent. Along much of the west coasts of the Americas and the East Coast of Asia we have recently discovered great fault planes shown by earthquakes as extending below the continental margins (Fig. 2-4). Scientists can tell from seismograms at what depth an earthquake takes place by comparing the time of its arrival at several seismographic stations. Going in from the Pacific coasts, the earthquakes are found to originate at greater and greater depths. Also, seaward of most Pacific coasts are long oceanic depressions that have the greatest depths in the world oceans (Fig. 2-4). These are called *trenches.* They are apparently the result of the sinking of the ocean crust beneath the advancing continents.

As the oceanic crust is pushed under the continents, it gets closer and closer to the hot interior. As a result, one would expect the rock to become molten. This is what happens, as we can see from the rows of volcanoes along most of the Pacific coasts. The volcanoes in the Andes inside the Peru-Chile Trench include some of the highest mountains in the world, up to elevations of 27,000 feet (8230 m). Less high are the island volcanoes inside the Aleutian Trench and volcanoes inside the Japan Trench.

Evidence for Migrating Continents and Oceans

We have seen that the ocean ridges, the rift valleys in their centers, the enormously deep trenches along the Pacific margins, and the rows of volcanoes along the adjoining continents and islands, all fit very nicely into the new sea floor-spreading hypothesis. More supporting evidence comes from ocean exploration. First, we have the deep-ocean drillings, a major undertaking supported mostly by the U.S.National Science Foundation and, more recently, receiving help from the Soviets, the British, the Swiss, the Canadians, and the Japanese. A decade ago, no one dreamed that we would soon have hundreds of holes drilled to depths of thousands of feet into the deep-ocean bottom. From the cores that have come up from these drillings we are now able to test some of the vital aspects of the new hypothesis. If the ocean crust has been migrating out from under the ridges, as in Fig. 2-3, the lavas should get older and older away from the ridge, and the ocean sediments that cover the lavas should become thicker and thicker away from the ridges, and should have increasingly older sediments in contact with the lavas. This is exactly what the deep-drilling operation has found (Figs. 2-3, 11-11, and Pl. 1). If, on the other hand, the Atlantic Ocean had been in existence before the Jurassic period, when the crack in the super-continent is supposed to have opened, we should drill into sediments that are definitely older than Jurassic before getting to the lava basement. To date, no older sediments have been discovered, either in the Atlantic or in any other ocean.

Using deep-diving vehicles, geologists have now made dives into one of the rift valleys in the Mid-Atlantic Ridge. In an expedition in 1973-74 called FAMOUS (French-American Mid-Ocean Undersea Study, Ballard et al., 1975), geologists accompanied the aquanauts and brought back clear evidence, including many photographs, that showed freshly opened cracks and blocks that had been pulled apart in very recent years. Also, they found freshly erupted lava flows that still were giving off considerable heat.

A more complicated type of evidence comes from measuring the magnetism of the rocks underlying the ocean floor. When molten rocks cool, the direction of magnetism of their grains become oriented toward the north magnetic pole. However, we have learned that during the past ages the earth's two magnetic poles have changed repeatedly at irregular intervals that average about a million years (for a further explanation, see Cox et al., 1967). By measuring the magnetism it is possible to tell whether the lava underlying the ocean cooled during a time when the north magnetic pole was in the Northern Hemisphere (called positive) or in the Southern Hemisphere (called negative). Therefore, if the sea floor-spreading hypothesis is correct, we should find bands of positive and negative magnetic anomalies extending parallel to the center line of the Mid-Atlantic Ridge. A sequence of positive and negative magnetic belts that have approximately the same sequence on the two sides of the ridge has been found in many crossings of the ridge (Pl. 1).* Vine and Matthews (1963), British geophysicists, made this discovery, although the existence of parallel belts of magnetism had been found previously by Mason and Raff (1961) from work off the California coast.

*Plate 1 is actually reproduced as an endpaper in this book. The two additional plate numbers therefore should be identified by changing Plate 1 to Plate 2, and Plate 2 to Plate 3.

The southern end of the Mid-Atlantic Ridge continues to the southeast around Africa and into the Indian Ocean (Fig. 2-5) where it splits, with one branch extending north into the Gulf of Aden and the Red Sea, the other bending around Australia and east into the Pacific Ocean, where it finally enters the Gulf of California and apparently goes under the North American continent.

Magnetometer lines across this world-encircling ridge have shown these same belts extending parallel to the other ridges and repeating in about the same sequence on the two sides. Although the comparisons do not match perfectly, there seems little doubt that these belts of alternating magnetism represent the sequence that has occurred in the earth's history. Since it is possible to date roughly the periods of positive and negative anomalies, it is also possible to tell by the width of each belt how fast the sea-floor spreading was progressing in past ages (Pl. 1). We find that spreading in the Atlantic was somewhat slower than in the Pacific Ocean: approximately one inch (2.5 cm) per year in the Atlantic and up to three inches (7.5 cm) per year in the Pacific. This may not seem fast, but with an average of two inches (5 cm) a year during the 150 million years since the opening of the great crack, the ocean floor could have moved 300 million inches or 5,000 miles (8,045 km). This would have allowed the de-

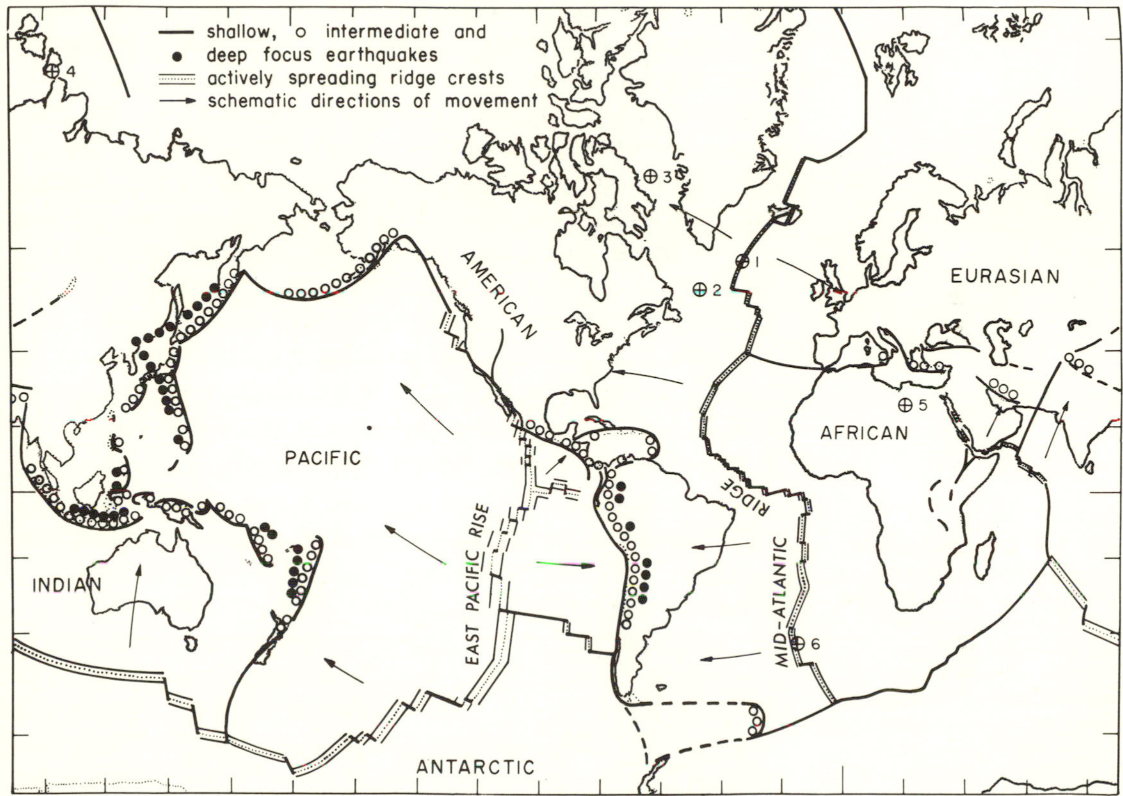

Fig. 2-5. The continuation of the Mid-Atlantic Ridge into the Indian and Pacific Oceans. The direction of drift of the various plates is shown by arrows. Note the offsets in the ridges produced by the fracture zones. See also Figs. 2-9B and 2-10.

velopment of the entire Atlantic and probably would have resulted in the subducting of every portion of the Pacific crust. So we should find no old crust under the oceans and, as previously noted, this is now established by ocean drillings.

In addition to these amazing recent proofs of crustal migrations over the surface of the earth, which have come from deep drillings, seismology, deep diving, and magnetism studies, we can now count much more on the old arguments used by Wegener for his somewhat similar hypothesis. The study of ancient fossils has convinced many geologists in recent years of the pre-Jurassic continental connections, but even more striking is the evidence for the 200 million-year-old Permo-Carboniferous glaciation. Extensive field work, especially by the American geologists Crowell and Frakes (1970) has shown the size of these ancient glaciers to be even larger than the Antarctica ice sheet. Furthermore, the study of the glaciated pavements with their scratches, called *striations*, indicates that glaciers in some places were coming out of what is now the ocean, and therefore can be explained best by the former connected land masses that have later been separated by sea-floor spreading (Fig. 2-6).

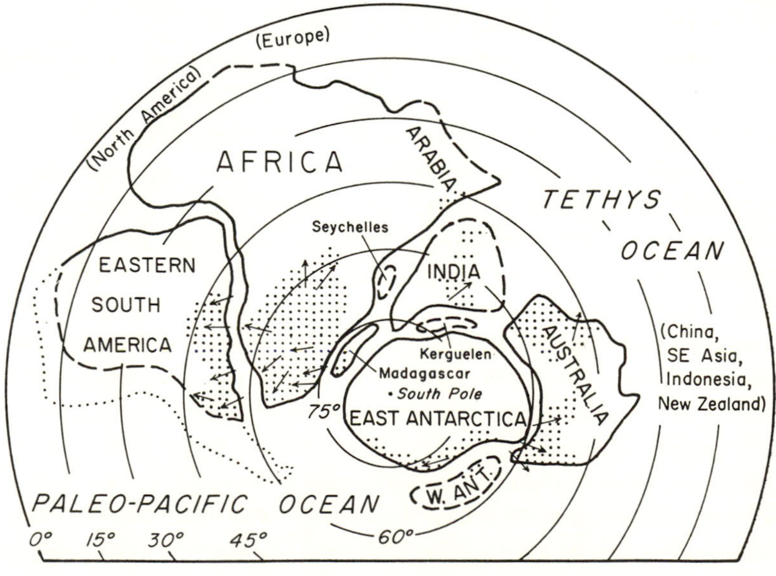

Fig. 2-6. The direction of giacial movements during the Permo-Carboniferous superimposed on the ancient continent of Pangea. If superimposed on the present continents, the glaciers would have to come out of the oceans. Also, the glaciers would have existed in tropical areas. From Hamilton and Krinsley, 1967.

Fig. 2-7. A portion of Chain and Romanche Fracture Zones in equatorial Atlantic. Contour interval of 200 fathoms except in irregular steep zones where indicated by shading. Arrows show flow of bottom water, so far as known.

Breaks in the Ocean Ridges and Transverse Fracture Zones

In the preceding discussion we have somewhat simplified the revolutionary new hypothesis of seafloor spread. This spread and the resulting continental sliding are now known to be far more complex than originally conceived. At first we thought of the great ocean ridges as curving but continuous. However, as exploration has advanced we have found that all the ridges are broken into many segments (Figs. 2-8, 2-9B, 11-6). This is shown clearly by the maps of the oceans constructed partly for the *National Geographic* by Bruce Heezen and Marie Tharp. The zones or lines that separate each of these offsets in the mid-ocean ridges are marked by escarpments, ridges, and elongate troughs (Fig. 2-7). Much to the surprise of marine geologists, it has been found that most of these transverse zones of marked relief can be traced for thousands of miles, and some of them cross the entire Atlantic and

Fig. 2-8. The direction of displacement along the fracture zones that results from lava rising along the crests of the mid-ocean ridges and being carried away from the ridges by sea-floor spreading. These are called *transform faults*. From Menard and Atwater, *Nature*, June, 1968.

more than half of the much wider Pacific (Fig. 11-6).

Because of these offsets, one would think that displacements have occurred along the ocean ridges, like those along the San Andreas Fault of California where objects like roads have shifted horizontally a few feet during earthquakes, but through our new methods of measuring the displacements during earthquakes, we have found that the motion taking place between the severed ridges indicates just the opposite direction (Fig. 2-8). This is nicely explained by the nature of seafloor spreading. If the cracks are being opened along the ridge crest and lava

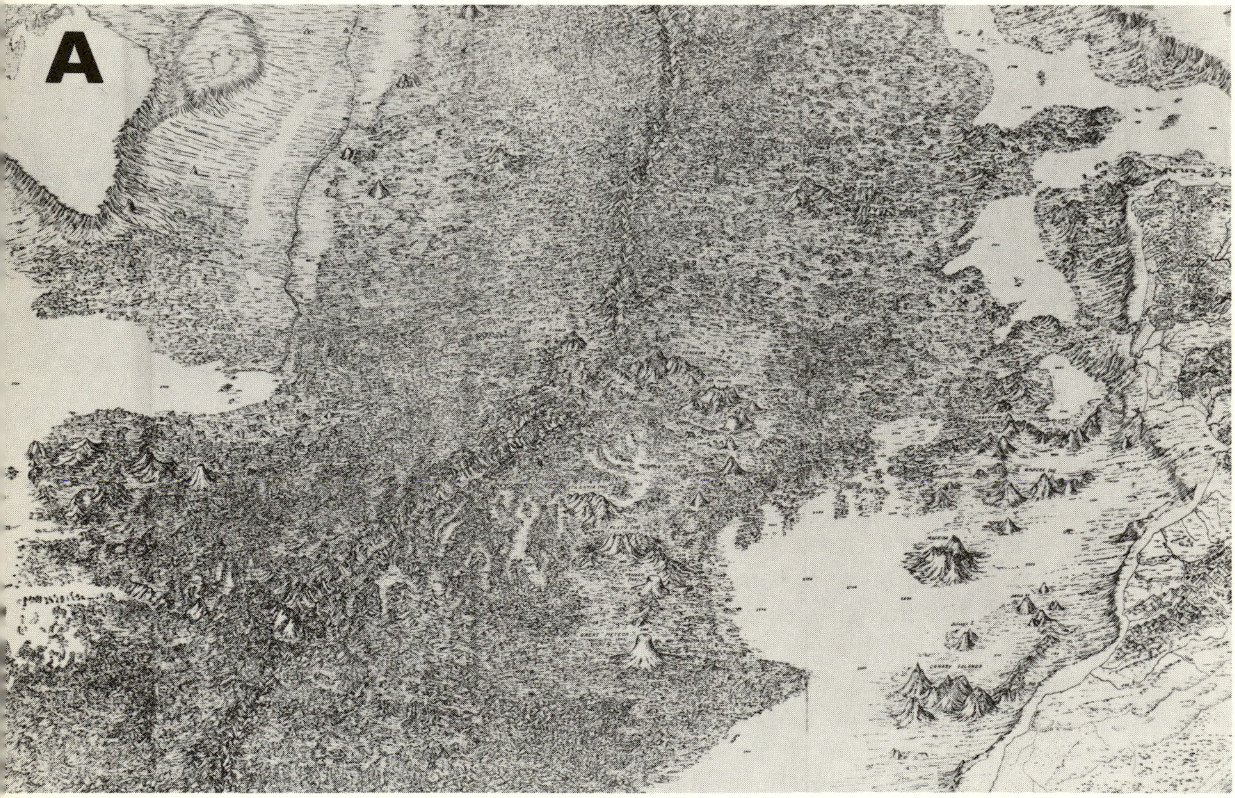

Fig. 2-9. Showing the considerable changes in the physiographic diagrams of the North Atlantic made by B. Heezen and M. Tharp (A) in 1959 and (B) (page 21) in 1968. The fracture zones were virtually unknown in 1959.

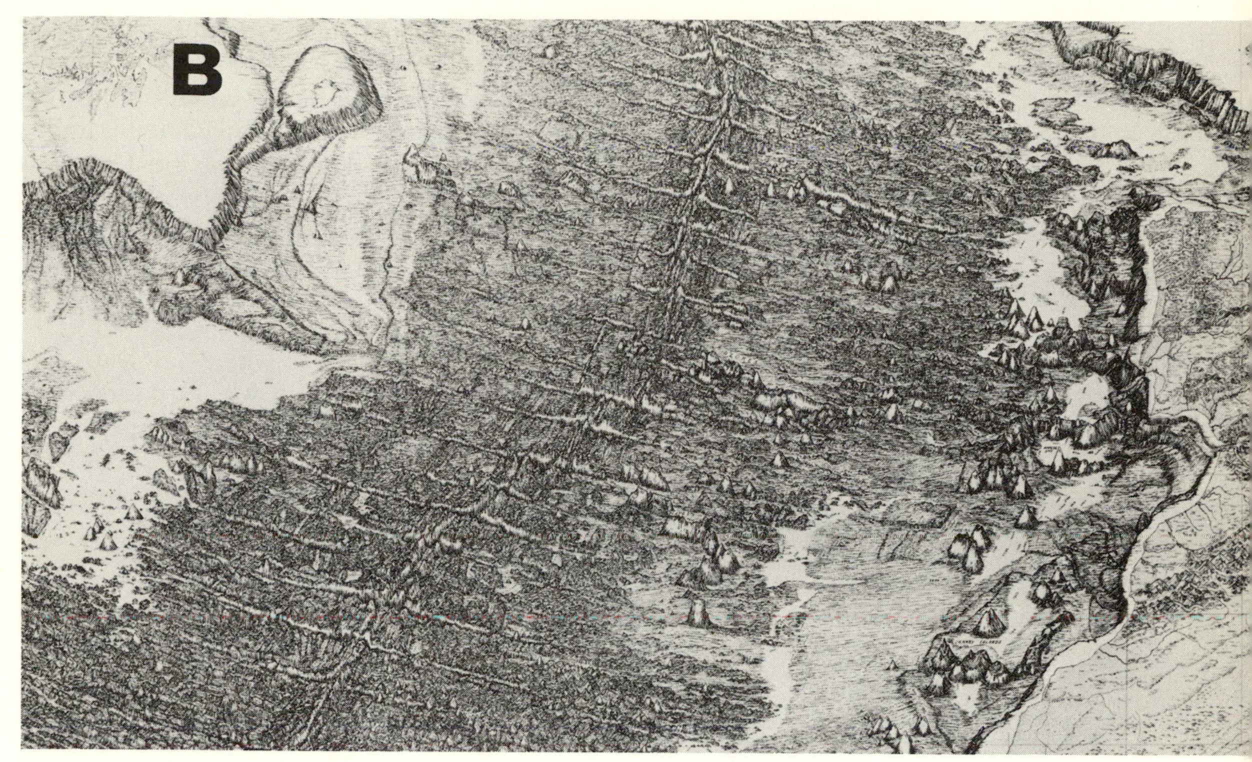

is rising along the cracks, as in Fig. 2-3, the displacement would be the same as has been found in earthquake studies. The change in appearance of the relief maps of the oceans during past decades is illustrated by comparing a portion of the Heezen and Tharp maps of the North Atlantic made in 1959 with those of 1968 (Fig. 2-9A and B). Equally large changes are seen in the maps of the Pacific. It is like comparing a map of the Rockies made at the time of the California Gold Rush with one of today.

Another complication in our simplified version of sea-floor spreading comes when one considers arcuate island chains, like the West Indies. Here we see a break in the continuity of the North and South American continents and, instead of finding a trench on the west side of the continents, we find that it lies to the northeast (Fig. 2-10). Furthermore, instead of the ocean crust sinking below the advancing west side of the continents, it is sinking below the northeast side. The same thing happens at the southern tip of South America, where the Scotia Arc bends eastward and extends in a curve down to Antarctica. These counter migrations are explained by movements of smaller plates in a direction opposite to the main trend that has opened up the Atlantic Ocean. Partly because of these complications, a new name has been coined for the sea-floor spreading hypothesis. It is now usually called *plate tectonics;* just another example of the host of new names that are coming into scientific usage.

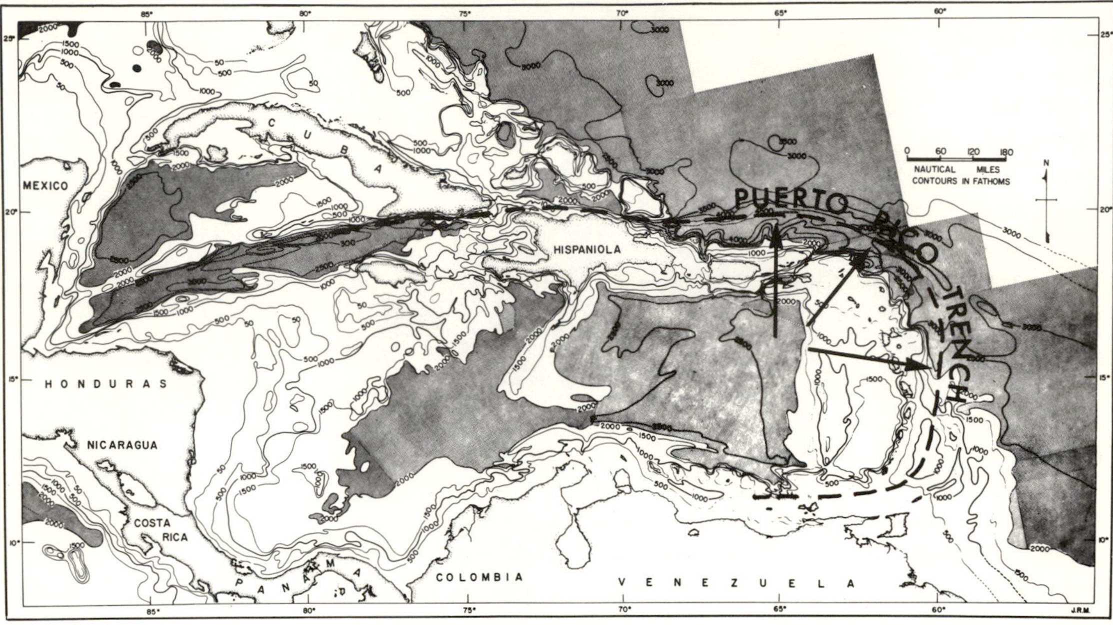

Fig. 2-10. The relief of the Caribbean and the West Indies where a plate has been pushed eastward between the west-moving North and South American plates. Note the Puerto Rican Trench on the north of Puerto Rico.

"Hot Spots" and the Oceanic Island Chains

Further evidence for the migratory nature of the earth's crust has come from a study of the ages of the volcanic island chains in the Pacific. Morgan (1972) introduced a new idea to account for some of these chains. Noting that the active volcanoes in the Hawaiian Islands are at the southeast end of the group and that, so far as is known, the volcanoes grow older to the northwest, Morgan suggested that the ocean crust is migrating northwestward over the underlying earth mantle and that the mantle has "hot spots" with sufficient heat to melt the overlying crust and cause breakthroughs of lavas that subsequently build volcanoes on the sea floor. The volcanoes then grow high enough to form islands. Presently, however, the oceanic crust has moved along on its northwesterly course and the mantle "hot spot" starts operating on a new portion of the crust farther to the southeast, so that a new volcano and its accompanying island is well developed. Meanwhile, the islands to the northwest start sinking due to cooling and due to their pressure on the ocean crust, although some of the sinking islands are kept near the surface by growth of coral reefs. As a result, we have many coral banks and atolls to the northwest of the Hawaiian Islands. A check of this hypothesis has been made by boring into Midway Atoll. Here, as reported by Ladd and others (1970) of the U.S. Geological Survey, lava was encountered below the coral. Its age is about 25 million years, considerably older than the lavas in the Hawaiian Islands.

The hot spot hypothesis has been applied to various other island chains in the Pacific, for example to the Tahiti-Bora Bora group. Some geologists think that the row of volcanoes that has been discovered in the deep ocean off New England originated in this way as well, but to date we lack confirmation. On the other hand, the Line Island chain south of Hawaii in the Pacific that trends northwest-southeast has had its volcanic rocks dated, and no consistent sequence is indicated. This suggests that not all lines of volcanoes are related to crust-migration over hot spots.

Present Status of the Migrating Earth Crust

In the preceding discussion we have outlined the pioneering hypothesis that puts the earth's crust on a journey, moving in massive plates over the earth's interior. The idea is so attractive and explains so many things so much better than the older ideas of permanent continents and orderly development of the earth's history that one is inclined to accept it. However, if one reads the extensive literature on the subject, one finds many geologists still in opposition (see for example Meyerhoff, 1970). Some of their arguments sound rather convincing, just as the arguments against Darwin's evolution theory often seemed convincing during the nineteenth century. The big question now is how fair are the arguments on both sides. Too many scientists are carried away by their own enthusiasm to such an extent that they tend to overlook the arguments on the other side. Despite its many attractive features, sea-floor spreading (or plate tectonics) has many things yet to explain. It is still difficult to understand how continents and ocean floors can wander so rapidly over the earth's interior. One should keep an open mind regarding this hypothesis to see what scientific progress will bring to us in the future.

References

Ballard, R. D., W. B. Bryan, J. R. Heirtzler, G. H. Keller, J. G. Moore, and Tj. van Andel, 1975. "Manned submersible observations in the FAMOUS area: Mid-Atlantic Ridge." *Science,* v. 190, no. 4210, pp. 103-16.

Bullard, E. C., 1969. "The origin of the oceans." *Sci. Amer.*, v. 221, no. 3, pp. 66-75.

Cox. A., G. B. Dalrymple, and R. R. Doell, 1967. "Reversals of the earth's magnetic field." *Sci.Amer.*, v. 216, no. 2, pp. 44-54.

Crowell, J. C., and L. A. Frakes, 1970. "Ancient Gondwana glaciations." *Second Gondwana Symposium, Proc. and Papers, South Africa,* pp. 469-76.

Dietz, R. S., 1961. "Continent and ocean basin evolution by spreading of the sea floor." *Nature,* v. 190, no. 4779, pp. 854-857.

Dietz, R. S., and J. C. Holden, 1970. "Reconstruction of Pangaea: breakup and dispersion of continents, Permian and present." *Jour. Geophy. Res.*, v. 75, no. 26, pp. 4939-4956.

Gutenberg, B., and C. F. Richter, 1954. *Seismicity of the Earth and Associated Phenomena.* Princeton Univ. Press, Princeton, N.J., 2nd Ed., 310 pp.

Hamilton, W., and D. Krinsley, 1967. "Upper Paleozoic glacial deposits of South Africa and southern Australia." *Geol. Soc. Amer. Bull.*, v. 78, no. 6, pp. 783-800.

Heezen, B. C., E. T. Bunce, J. B. Hersey, and M. Tharp, 1964. "Chain and Romanche Fracture Zones." *Deep-Sea Res.*, v. 11, no. 1, pp. 11-33.

Hess, H. H., 1962. "History of ocean basins." In *Petrologic Studies: A Volume in Honor of A. F. Buddington.* A. E. J. Engel et al., eds., Geol. Soc. Amer., Boulder, Colo., pp. 599-620.

Ladd, H. S., J. I. Tracey, Jr., and A. G. Gross, 1970. "Deep drilling on Midway Atoll." *U.S. Geol. Surv. Prof. Paper 680-A,* pp. A1-A22.

Mason, R. G., and A. D. Raff, 1961. "Magnetic survey off the west coast of North America. 32° N latitude to 42° N latitude." *Geol. Soc. Amer. Bull.*, v. 72, no. 8, pp. 1259-1266.

Menard, H. W., and T. M. Atwater, 1968. "Changes in direction of sea floor spreading." *Nature,* v. 219, no. 5153, pp. 463-467.

Meyerhoff, A. A., 1970. "Continental drift, implications of paleomagentic studies, meteorology, physical oceanography, and climatology." *Jour. Geol.*, v. 78, no. 1, pp. 1-51.

Morgan, W. J., 1972 "Deep mantle convection plumes and plate motions." *Amer. Assoc. Petrol. Geol. Bull.*, v. 56, no. 2, pp. 203-213.

Revelle, R. R., 1955. "On the history of the oceans." *Jour. Mar. Res.*, v. 14, no. 4, pp. 446-461.

Vine, F. J., and H. H. Hess, 1970. "Sea-floor spreading." In *The Sea,* Vol. 4, Pt. 2, A. E. Maxwell, ed., Wiley-Interscience, N.Y., pp. 587-622.

Vine, F. J. and D. H. Matthews, 1963. "Magnetic anomalies over oceanic ridges." *Nature*, v. 199, no. 2, pp. 947-949.

Wegener, A., 1924. *The Origin of Continents and Oceans.* Eng. transl., Dutton N.Y., 212 pp.

Suggested Supplementary Reading

Alexander, T., 1975. "A revolution called plate tectonics has given us a whole new earth." *Smithsonian,* v. 5, no. 10, pp. 30-40.

Alexander, T., 1975. "Plate tectonics has a lot to tell us about the present and future earth." *Smithsonian,* v. 5, no. 11, pp. 38-47.

Degens, E. T., and D. A. Ross, 1970. "The Red Sea hot brines." *Sci. Amer.*, v. 222, no. 4, pp. 32-42.

Dewey, J. F., 1972. "Plate tectonics." *Sci. Amer.*, v. 226, no. 5, pp. 56-68.

Dietz, R. S., 1972. "Geosynclines, mountains and continent-building." *Sci. Amer.*, v. 226, no. 3, pp. 30-38.

Dietz, R. S., and J. C. Holden, 1970. "The breakup of Pangaea." *Sci. Amer.*, v. 223, no. 4, pp. 30-41.

Hallam, A., 1972. "Continental drift and the fossil record." *Sci. Amer.*, v. 227, no. 5, pp. 56-66.

Heezen, B. C., 1960. "The rift in the ocean floor." *Sci. Amer.*, v. 203, no. 4, pp. 99-110.

Heezen, B. C., M. Tharp, and M. Ewing, 1959. "The Floors of the Oceans I. The North Atlantic. Text to Accompany the Physiographic Diagram of the North Atlantic." *Geol. Soc. Amer.*, spec. pap. 65. (And the Revised 1968 Physiographic Diagram by Heezen and Tharp; Williams and Heintz Map Corp., Wash., D.C.)

Hsü, K. J. 1972. "When the Mediterranean dried up." *Sci. Amer.*, v. 227, no. 6, pp. 26-36.

Hurley, P. M. 1968. "The confirmation of continental drift." *Sci. Amer.*, v. 218, no. 4, pp. 52-64.

James. D. E., 1973. "The evolution of the Andes." *Sci. Amer.*, v. 229, no. 2, pp. 60-69.

Kurtén, B., 1969. "Continental drift and evolution." *Sci. Amer.*, v. 220, no. 3, pp. 54-64.

Menard, H. W., 1961. "The East Pacific Rise." *Sci. Amer.*, v. 205, no. 6, pp. 52-61.

Menard, H. W., 1969. "The deep-ocean floor." *Sci. Amer.*, v. 221, no. 3, pp. 126-142.

Orowan, E., 1969. "The origin of the oceanic ridges." *Sci. Amer.*, v. 221, no. 5, pp. 102-119.

Scientific American, 1969. *The Ocean.* W. H. Freeman and Co., San Francisco. Chs. 2 and 5.

Toksöz, M. N., 1975. "The subduction of the lithoshere." *Sci. Amer.*, v. 233, no. 5, pp. 88-98.

Wilson, J. T., 1963. "Continental drift." *Sci. Amer.*, v. 208, no. 4, pp. 86-100.

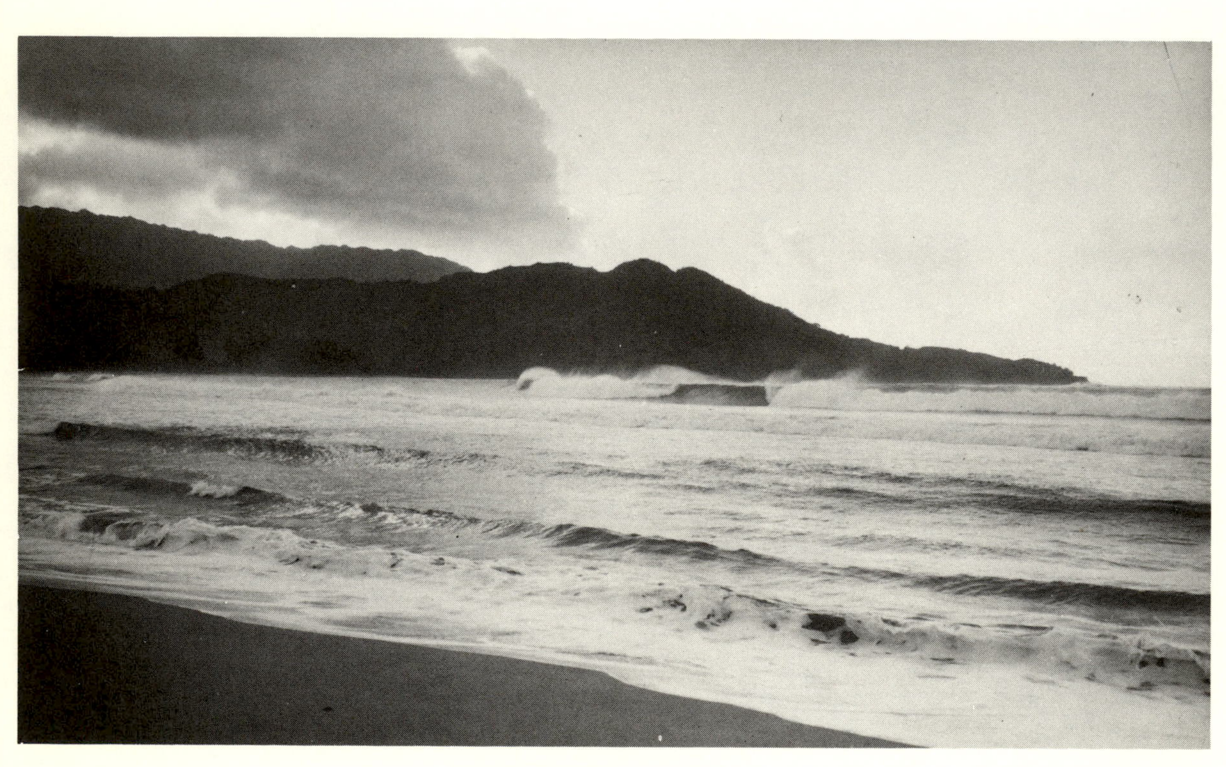

Waves and currents-the unsteady sea

Chapter 3

Cruise advertisements usually show a ship moving through a sea flat as a diningroom table, but in reality a passenger with a squeamish stomach is much more impressed with the irregularities of the sea surface. Small boats in the open sea are almost constantly rolling and pitching as a result of the waves and currents, and even large steamers are rarely without motion.

Swimmers look eagerly at wave forecasts for two very different reasons. The surfers are looking for large waves that produce big curling breakers with an edge, which they ride diagonally toward the shore on their surf boards, but the less sports-minded swimmer hopes for small waves that are pleasanter and less dangerous. Also, the small waves do not roil the sea water; hence greatly improve the visibility, allowing the swimmers with face masks to look at the beautiful fish and the plants and encrusting animals that grow on the bottom.

Nature of Waves

Most waves are due to winds exerting frictional stress on the sea surface. The more unusual catastrophic waves caused by movements of the sea floor at the time of earthquakes, submarine volcanic activity, or by great landslides will be discussed in the next chapter.

When the wind moves along the sea surface it first produces ripples, but as the wind continues, these ripples combine to make what we call *wind waves.* The nature of waves is illustrated in Figure 3-1A. The surface alternately moves up and down in an orbit, forming crests and troughs. The time between passage of successive crests (or troughs) is called the *wave period.* The distance between two crests is called the *wave length,* and the vertical distance between trough and crest is called *wave height.* Wind-wave periods may be as long as 20 seconds in the open ocean, and wind-wave lengths may be as much as 1,000 feet (305 m). The waves with the longest periods have the greatest lengths.

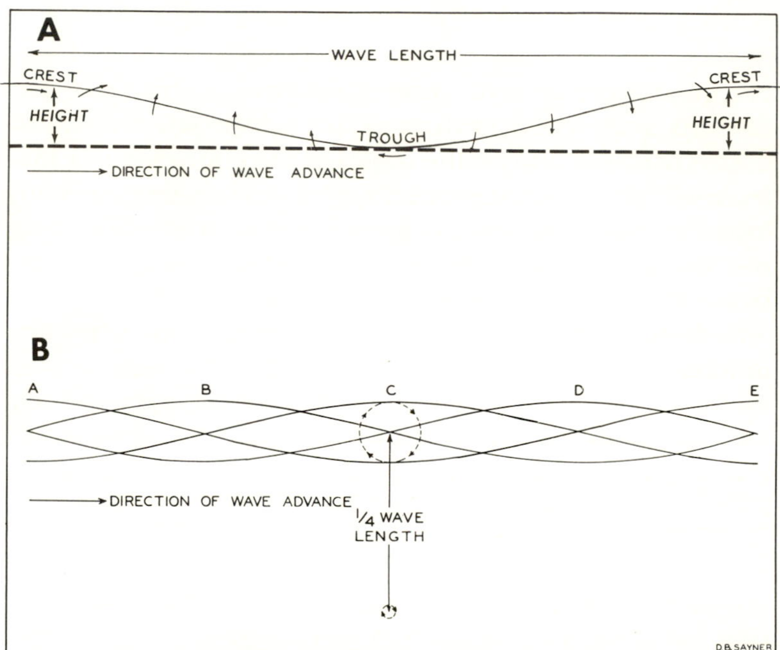

Fig. 3-1A. Terms applied to waves.

Fig. 3-1B. Variation of waves with depth.

By putting a drop of colored water inside the glass side of a tank in which waves have been set up by a paddle, one can see that the water particles make a somewhat circular orbit, but actually move forward like a valve cap on a tire that is rolled along the ground (Fig. 3-1B). The advance is usually in the direction in which the wind is blowing, and the wave form is steeper on the advancing front. Water particles move much slower when the wave crest appears to be advancing, as you can again see from watching the particle of colored water rotate along the glass side of the tank. In the open ocean, the waves commonly advance at rates of about 10 to 60 miles (16 to 97 km) per hour. In general, the faster the rates, the greater the wave length.

Perhaps the most interesting feature of sea waves is their ability to advance from the area where they are developed by winds. We have discovered large waves produced by great storms thousands of miles away from their generating area. A great storm near Antarctica may send waves of large size to the California coast and a storm off the coast of Morocco may send waves across the Atlantic to the East Coast of the United States. Waves sent out from a distant storm are called *swell*. They differ from the waves in the storm-generating area by having longer periods, greater length, much smoother surfaces, and in having approximately the same slope on both sides. Surfers greatly prefer swell waves to wind waves. For that reason, very good surfing is found on the Southern California coast and on the coasts of the Hawaiian Islands, from which the great storms are usually distant. In both localities, the large swell come from the northwest during the winter and from the southwest during the summer, because of the northern winter storms in the North Pacific and the southern winter storms in the high southern latitudes.

Shoaling and Breaking Waves

As waves approach a coast they tend to change their direction to conform to the bottom slope and to the coastline. As the waves advance into shoal water (where the depth becomes less than about half the wave length), the wave crests begin to get more closely spaced, develop greater height, and slow their rate of advance, although the period remains the same. Where the bottom is irregular, the waves will first be affected over the shoals so that individual waves will advance faster over the deep areas. For example, the portion of a wave following along the floor of a sea valley will get ahead of the portions on either side (Fig. 3-2A and B). As a result, the wave will throw some of its energy toward the shoal portion of the sea floor, and the wave will be reduced in height over the valley and increased on the flanks. Fishermen, as a result,

Fig. 3-2. Showing how wave crests are slowed in moving up a ridge (A) and accelerated in moving up a valley (B). This increases height on ridges and decreases height at heads of valleys.

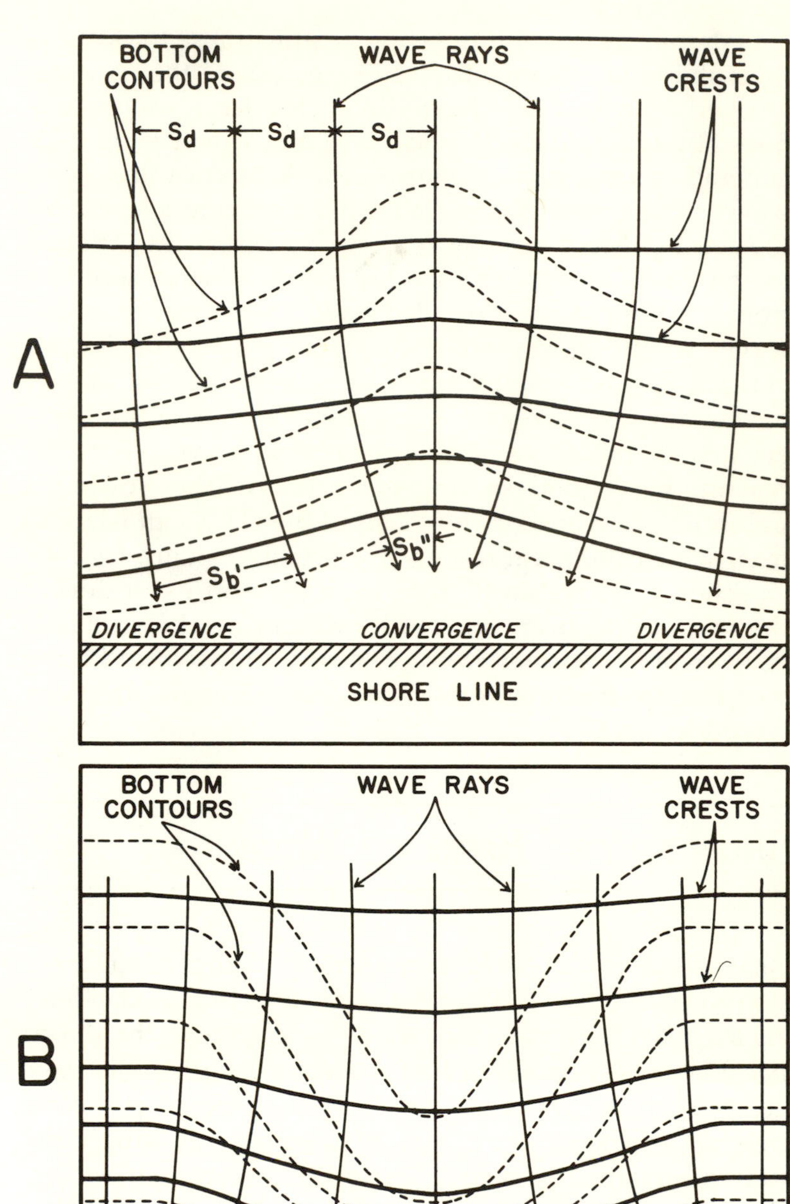

choose a location for launching their boats at the head of a sea valley, and surfers choose the adjacent areas for the large waves. Many examples of this can be seen along the California coast, and in western Europe the fishing towns of Nazare in Portugal and Cap Breton in southwestern France are located at the heads of submarine canyons.

Fig. 3-3A. A spilling breaker during a strong wind.

As the waves finally get into very shoal water they become unstable, and at a depth of about 1.3 times their height the wave crest falls forward or breaks, and the water rushes up the shore. This is called *uprush.* Uprush is followed by *backwash,* which returns the water that has not seeped into the beach. The nature of the breaker depends on whether the waves that come into the shore are wind waves or swell. Wind waves produce spilling breakers comparable somewhat to stream rapids (Fig. 3-3A); whereas swell waves have plunging breakers that look like

Fig. 3-3B. Hollow-front plunging breaker developing from swell.

a waterfall (Fig. 3-3B). The plunging breakers with their hollow fronts are the most dangerous, because they may fall on top of a swimmer and push his head into the bottom or even break his neck. Expert surfers are often able to ride their boards into the tunnel under the plunging breakers and may even emerge on the far side still riding their boards, but such a ride is perilous, to put it mildly.

Longshore and Rip Currents

If the waves are breaking near the shore, you can throw a stick out from the beach beyond the breakers, then watch the progress of the floating stick. In addition to moving toward the shore as it rises on each wave crest and moving offshore to a smaller degree in the wave troughs, the stick will probably move parallel to the beach. These flows, called *longshore currents,* may move your stick as fast as you can walk. If you see the direction in which the waves appear to be approaching the beach, you will usually find that the stick is progressing in the same direction. This is important to remember if you leave your swim suit and towel on the beach and go for a skinny dip. While jumping up and down in the waves, you may unknowingly be carried along the shore for such a distance that you don't find your clothes when you do come ashore.

In describing the motion of waves, we have seen that water particles have a net movement in the same direction as the crests' advance but at a very much slower speed. Because the crests ordinarily move toward the beach, the tendency is to pile up water along the shore. However, this could not continue without flooding the coast, so actually the water must be moved along the coast until it gets to a point where it can return seaward. Geologists used to think, not too many years ago, that the water pushed shoreward was returned as an *undertow.* People will swear to you that they have been caught by the undertow and carried out along the bottom until they finally escaped to the surface and saved their lives. Some years ago, I had my interest aroused in this undertow idea by a lifeguard who patrolled the beach at Santa Monica, California. He said, "Undertow is a myth. You come with me and I'll show you what really happens to return that water to the sea." We entered the waves where there was a strong longshore drift. Following this, we found after a short distance that the current had changed direction and was bending seaward where its speed increased. Trying to swim back toward the beach we found it was too strong to make any progress. The lifeguard said "Either let it carry you out or swim along parallel to the shore and you will get out of this current." It was a relief to find that both of these methods worked.

The phenomenon we encountered was what is popularly called a *rip tide* or, more scientifically, a *rip current,*

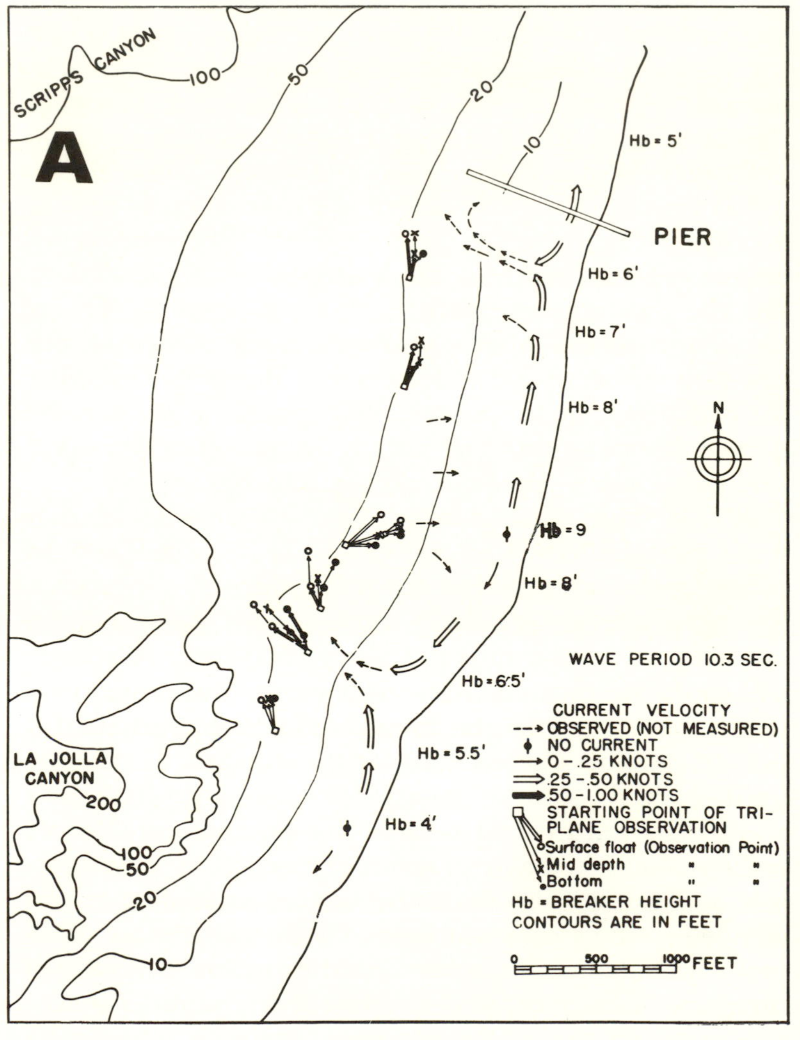

Fig. 3-4A. Diagram showing longshore currents as related to rip currents at Scripps Institution of Oceanography, La Jolla, California. Note canyons.

because the tide is not involved. Rip currents usually form a pattern, as in Figure 3-4A, moving out from the beach along a channel perhaps 20 feet wide, and then 100 or more feet from shore spreading out with a slower speed. Finally the water starts to return to the beach, as in Figure 3-4B. Along patrolled beaches, the rips are often indicated

Fig. 3-4B. Photograph of a rip current outlined by seaward-moving foam lines.

by signs warning swimmers to keep out. You can usually see rips, however, by the streaks of foam that extend out through a gap in the breakers and the clouds of dirty water in the current and on the outside. Looking down from a bluff, they are easy to detect. However, if you do get caught in one, don't try to swim against it and wear yourself out. That is how many people drown. Swim to one side and, if that does not work, try the other side. Also, make use of every large wave crest by riding it in toward the shore. Remember that keeping afloat requires almost no energy, so if you relax and take your time, you will let the circular motion of the current return you to the shore. One more piece of advice—unless you are an excellent swimmer, keep out of the water when there are large breakers that cut the horizon above your height of eye when you are standing at the water's edge. Strong rips are rarely developed except during times of large waves.

Rip currents often occur along the side of a point or projecting sand spit where the waves have been driving the water along the coast until they come to an obstruction where the current turns seaward. Another dangerous place where one can be carried seaward by a special type of rip is at a channel through a coral reef. If the waves have been breaking over the reef and building up the water in the lagoon on the inside, there may be a sudden outflow of water up to several miles per hour through the channel. While swimming at Midway Island, I was caught in one of these and saved myself by holding onto a small coral head until the strong surge had abated. If you let yourself be carried out through the reef channel to the open sea, you may have to wait to be picked up by a passing boat or have the painful experience of trying to ride a breaker up onto the rough coral reef, a bad choice (see Chapter 9 on the coral reefs).

Tides and Tidal Currents

Tides are the result of the attraction of the moon and, to a lesser extent, of the sun, on large bodies of water. If you pull someone sideways by the arm, both of his arms will move away from his body, one by direct pull and the other by inertia. That is approximately what happens to form the ocean tides, except that both the moon and the sun are producing the pulls and they may be working together or against each other. Together means on the same side or on directly opposite sides, the latter being comparable to having someone pull your left arm while someone else pulls your right arm. Moon and sun working against each other would be like someone pulling you by the shirt front while someone else pulled one of your arms. The highest and lowest tides occur when the sun is either on the same side of the earth as the moon (new moon) or on opposite sides (full moon), and the tides with the smallest rise and fall occur when the moon is at right angles to the sun (first or third quarter).

Now look for a surprise, as I found when I was becoming interested in the oceans. You see the moon is full and overhead, and you go to the beach expecting the tide to be dead high along the shore, but you find that the tide is low, embarrassing if, as in my case, you had been teaching elementary geology students that the tide would be high under such circumstances. What was wrong? The tide is not as simple as the arm experiment would indicate. Actually, the attraction of the moon and sun causes currents to move up and down the coasts on the two sides of the ocean, rather than just straight in and out. You look at a tide table and you will find, for example, that Boston has a low tide, say, at 1029 and Charleston, South Carolina at 0702. Similarly, San Diego, has a low at 0807 and San Francisco at 1026. Another thing you will find is that areas at the heads of long inlets, like the Bay of Fundy and the Narrows of the English Channel, have very high tides, and they occur at times much later than at the mouths of these inlets. If the tide is much retarded, it may come in as a wave, called a *tidal bore* (Fig. 3-5). In general, the tides have the smallest ranges along the shores of mid-ocean islands, like Hawaii and Tahiti, and in inland seas, like the Mediterranean, the Caribbean, or the Gulf of Mexico.

Currents due to tides may be very strong, particularly in narrow straits where the high tide on one side coincides with the low tide on the other. For example, between Vancouver Island and the mainland of British Columbia, Seymour Narrows used to have a tidal current that flowed as fast as 16 knots past a slightly submerged rock in the center, so vessels had to move through this strait when the tide was relatively high and flowing in the right direction. By blasting out the large submerged rock in the narrow part of the strait, the passage was considerably deepened. As a result, now the current is slightly reduced and the passage is much less treacherous. Strong tides occur at the narrow entrances to larger bays (like the Golden Gate at San Francisco) where the tide flows at speeds up to 6 knots. The internal seas of Japan also have strong currents (up to 11 knots at their entrances). This fact has allowed American submarines to get into these Japanese embayments during World War II because anti-submarine nets were difficult to emplace and the noise of the tide made detection almost impossible.

Ocean Currents, the World's Greatest Rivers

Benjamin Franklin, with his interest in natural phenomena, was the first man to make a concerted study of the Gulf Stream. His collection of notes from ship captains showed him the importance of this great ocean river that flows north along the southeast coast of the United States as far as Cape Hatteras, where it swings seaward and finally crosses the Atlantic, carrying relatively warm water to the northwest coast of Europe. This is what makes an equable climate in Europe at the same high latitudes where frigid weather is prevalent in Labrador, on the other side of the ocean.

Fig. 3-5. Tidal bore advancing up Turnagain Inlet, Alaska, near Anchorage.

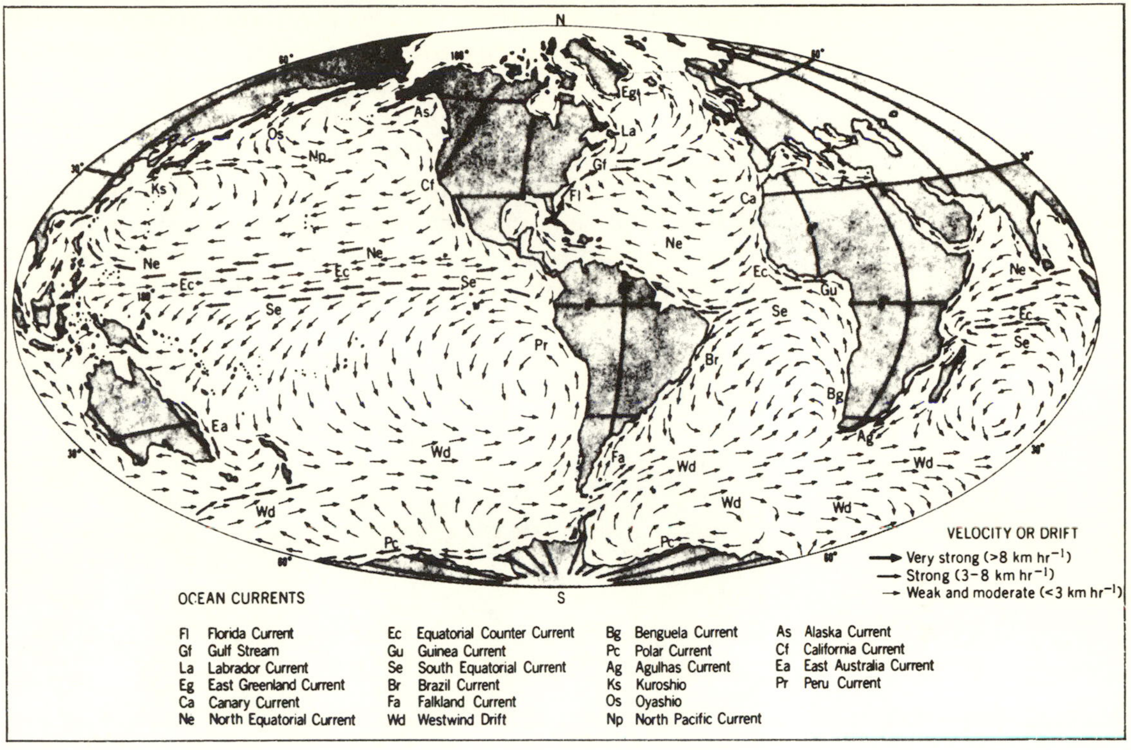

Fig. 3-6. The major ocean currents. Note the directions change to the right in the Northern Hemisphere and to the left in the Southern Hemisphere. After Shepard, 1973, Figure 3-15.

Because of its importance to navigation, especially for slow sailing vessels, the navies of the world developed considerable knowledge of the Gulf Stream and of the other major ocean surface currents. Thus our maps (Fig. 3-6) of the surface currents today look about the same as they did a hundred years ago. The major ocean currents have rotary courses, turning clockwise in the Northern Hemisphere and counter-clockwise in the Southern. They are strongest on the west side of the oceans in the Northern Hemisphere and on the east side in the Southern Hemisphere. Recent measurements of their velocities have shown speeds up to 8 knots, instead of the 4 knots thought likely in early days. Oceanographers have discovered migrating eddies on the outer margin of the Gulf Stream along the East Coast. The relationship of relatively deep cold currents supplied from the Norwegian Sea to warm currents has also been well established. In the North Atlantic, the cold currents follow down the east coast of North America, moving under the warm currents; in the Southern Hemisphere, the cold currents follow up the west coast of Africa, where they also flow beneath the warm currents. The most spectacular example of a north-moving cold current is the Humboldt Current that at times flows up western South America as far as the Equator, so that one may actually encounter 60°F (15°C) water at the Galapagos Islands at 0° latitude.

Ocean currents might be thought of as having little to do with marine geology if, as was supposed, they are mainly rivers of ocean water flowing well above the bottom except where they come in close to the coast. However, here

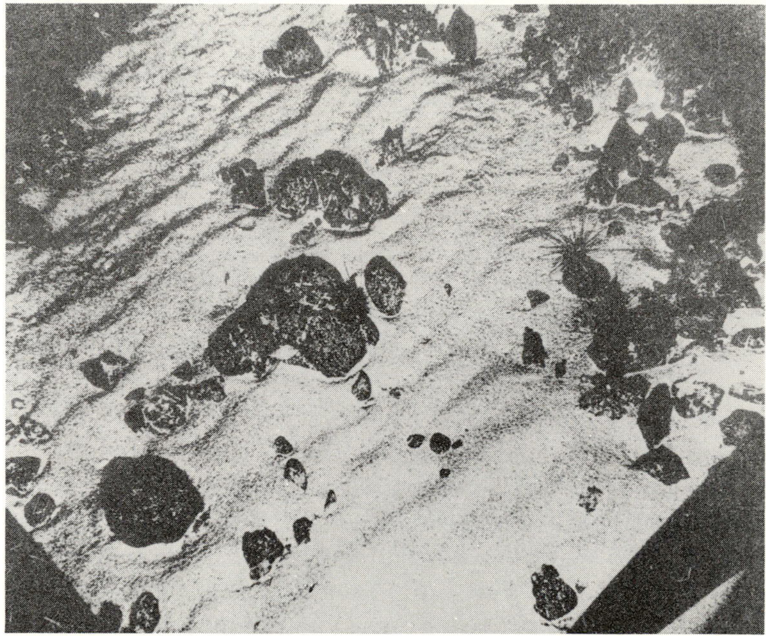

Fig. 3-7. Ripple marks on the deep-ocean floor. The rocks are manganese nodules. See Chapter 11.

again we have had a revolution in our ideas. Due to the phenomenal success of such marine geologists as Bruce Heezen and Charles Hollister (1971) in photographing the deep-ocean bottom, particularly in the Atlantic, we have found that there are currents of sufficient strength to produce ripple marks (Fig. 3-7), underwater dunes, and even sea-bottom gouges, all in water more than two miles (3.2 km) deep.

These currents along the east coast of North America are produced by the cold water that comes from the overflow of the Norwegian Sea and sinks below the Gulf Stream. In fact, the deep-ocean floor may have relatively strong currents in many places. Where pipes have been driven into the bottom to obtain cores, it is often possible to use the small fossils (such as foraminfera and radiolarians) that are embedded in the sediments to determine the geological age when the deposits were formed. William Riedel (1971) began finding that even in many short cores there were sediment layers containing fossils of animals that were known to have lived many millions of years in the past. Because sediment is being carried to all parts of the ocean by the moving surface waters, one has to conclude that these old sediments were probably preserved at or near the bottom due to currents of sufficient strength to carry away the newly introduced sediments. The Deep-Sea Drilling Project, with its hundreds of borings in the oceans all over the world, has completely confirmed Riedel's findings from his short cores. In many places, sediments of old geological ages are missing, showing that at various times in the past currents have prevented deposition and have even cut away some of the old deposits.

Turbidity Currents

Reginald Daly (1936), having read about the sediment-laden currents that flow down the slopes of newly created lakes (such as Lake Mead), suggested that ocean water could also become sufficiently heavy, due to sediment being stirred up by the waves or by landslides, that this water could flow rapidly down submarine slopes and even cut great canyons in the bottom. Bruce Heezen and Maurice Ewing (1952) brought great support to Daly's idea by calling attention to the sequence of breaks in cables that occurred off the Grand Banks in 1929 at the time of the world-shaking Grand Banks earthquake (Fig. 3-8). Using time and distances between cable breaks, they concluded that the turbidity currents set up by this great earthquake actually moved down the slopes as fast as 60 miles (97 km) per hour. To some of the rest of us this has seemed faster than the facts could justify, but the cables certainly did break in sequence, and overlooking the first two breaks that may have been started by the earthquake itself, there appears to be good grounds for a rate of advance down the slopes of about 17 miles (27 km) per hour. Such a current would produce active cutting of the sea floor.

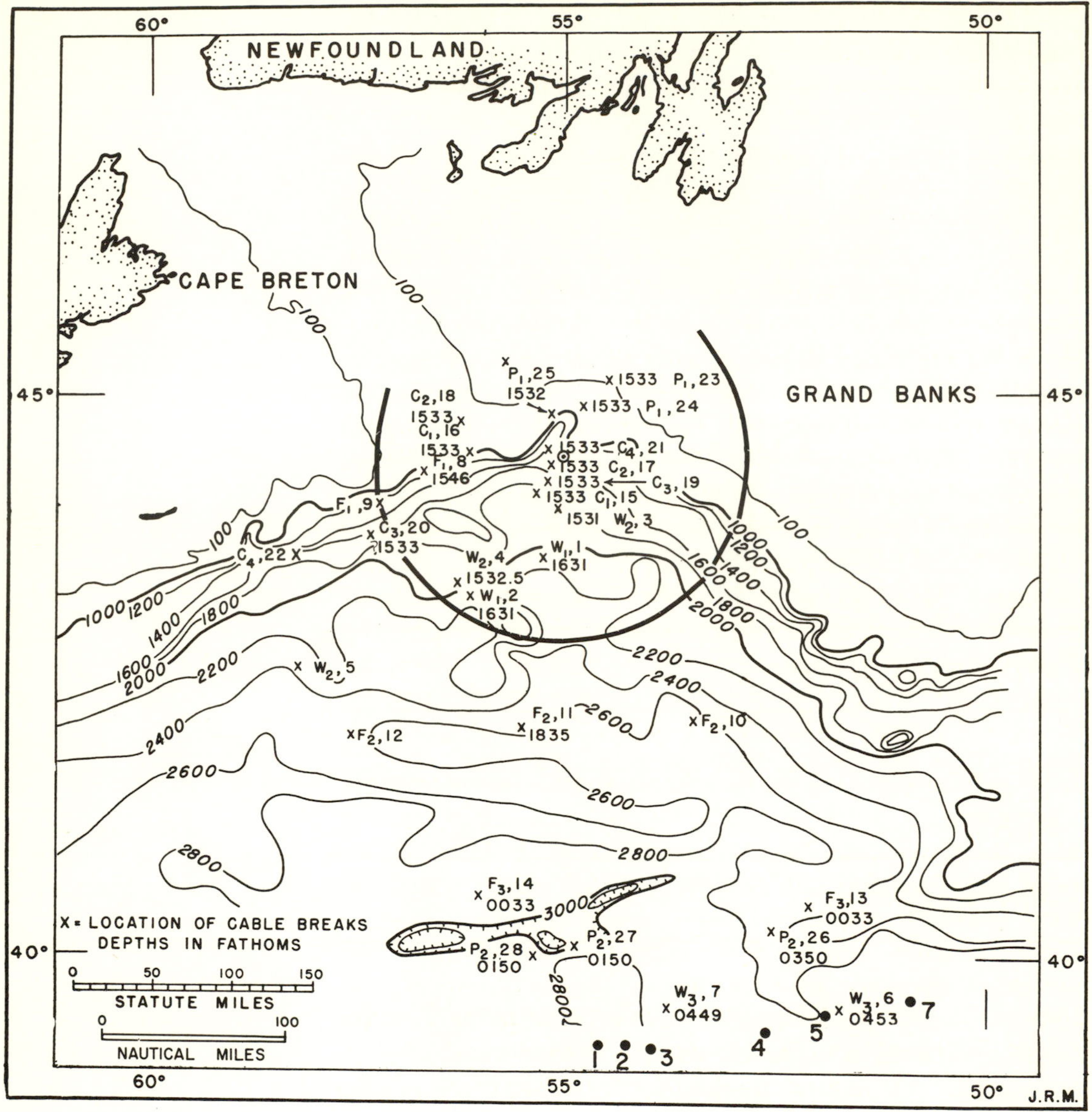

Fig. 3-8. Showing the times of cable breaks, apparently due to turbidity currents at the time of the Grand Banks earthquake. In the zone within the circle, breaks were probably due to slumping at or shortly after the earthquake. The numbers refer to samples of the bottom.

In my own work of recent years, we have been putting current meters out into the valleys of the sea floor to measure the currents. One great difficulty of the project has been that turbidity currents occur from time to time and carry away our expensive meters, making it impossible to determine the speeds of these currents. However, we at least know that these turbidity currents were started, not by earthquakes, but by storms with strong onshore winds. Apparently the storm waves stir the bottom sufficiently to allow sediment to be mixed into the water to start a strong flow down the valleys.

Internal Waves

From our (Shepard et al., 1974) studies of submarine canyons we find that the currents move alternately up and down the canyon floors, sometimes as fast as one mile (1.6 km) per hour. By obtaining a pattern of alternations of current velocity at two or more stations in a valley, it is possible to determine that these alternations are advancing along the valley floors, usually upvalley. For example, a relatively fast flow on the current-meter record at one point can be found repeated, say, an hour later at a station upvalley. Some time before World War I, oceanographers discovered that temperature zones were rising and falling below the surface without any corresponding surface waves. These subsurface phenomena are now called *internal waves.* More recently, after World War II, oceanographers found that these internal waves advance across the continental shelf toward the shore (LaFond, 1966). By comparison, it seems evident that the alternating currents in submarine valleys are also a type of internal wave, although they are often closely related to the tides.

References

Daly, R. A., 1936. "Origin of submarine canyons." *Amer. Jour. Sci.,* ser. 5, v. 31, no. 186, pp. 401-420.

Heezen, B. C., and M. Ewing, 1952. "Turbidity currents and submarine slumps, and the Grand Banks earthquake." *Amer. Jour. Sci.,* v. 250, pp. 849-873.

Heezen, B. C., and C. Hollister, 1971. *The Face of the Deep.* Oxford University. Press, N.Y. and London, 659 pp.

LaFond, E. C., 1966. "Internal waves." In *Encyclopedia of Oceanography,* R. W. Fairbridge, ed., Reinhold Publ. Corp., N.Y., pp. 402-408.

Riedel, W. R., 1971. "The occurrence of the Pre-Quaternary Radioloria in deep-sea sediments." In *Micropaleontology of Oceans,* B. M. Funnell and W. R. Riedel, eds., Cambridge University Press, pp. 567-594.

Shepard, F. P., 1973. *Submarine Geology.* 3rd Ed. Harper & Row, N.Y., Fig. 11-18.

Shepard, F. P. and D. L. Inman, 1950. "Nearshore water circulation related to bottom topography and wave refraction." *Trans. Amer. Geophys. Union,* v. 31, no. 2, pp. 196-212.

Shepard, F. P., N. F. Marshall, and P. A. McLoughlin, 1974. "Currents in submarine canyons." *Deep-Sea Res.,* v. 21, no. 9, pp. 691-706.

Suggested Supplementary Reading

Bascom, W., 1960. "Beaches." *Sci. Amer.,* v. 203, no. 2, pp. 80-94.

Bascom, W., 1964. *Waves and Beaches, the Dynamics of the Ocean Surface.* Anchor Books, Doubleday & Co., Inc., Garden City, N.Y. Chs. 1-8.

Chapin, H., and F. G. W. Smith, 1952. *The Ocean River.* Charles Scribner's Sons, N.Y., 325 pp.

King, C. A. M., 1972. *Beaches and Coasts.* 2nd Ed., St. Martin's Press, N.Y. Chs. 3-5.

Lissau, S., 1976. "Beach Currents." *Oceans,* v. 9, no. 4, pp. 14-21.

Van Dorn, W. G., 1974. *Oceanography and Seamanship.* Dodd, Mead & Co., N.Y. Chs. 12-17.

Catastrophic waves

Chapter 4

Many notorious disasters have been the result of huge sea waves, popularly called *tidal waves*. But, as in the case of rip tides, tidal waves have very little to do with the tides. The large sea waves that sometimes accompany earthquakes or submarine volcanic activity are now generally referred to as *tsunamis*. *Seismic sea waves* is a name also used by some scientists for the type caused by earthquakes. The huge engulfing waves that often accompany hurricanes are appropriately called *storm surges*. A much less common type of sea wave but one of great potential destructive power is due to landslides where rock masses break away from sea cliffs, catapulting huge masses of material into the sea or into a lake. These are called *landslide surges*.

Tsunamis and their Prediction

When the sea bottom suddenly moves up or down, causing an earthquake, water above the moving block is displaced and in turn sets up a train of waves. For thousands of years, the Japanese who live along the coastal lowlands have been in terror of these waves, which from time to time have destroyed their homes and drowned thousands of people. The Japanese coined the word *tsunami*, meaning "high water in a harbor," because these waves with very long periods do come into harbors with disastrous effects. The movements of the sea bottom off Japan often occur near population centers, so there is generally a short time interval between the arrival of the fast-traveling earthquake tremors and the arrival of the large sea waves. However, in the Hawaiian Islands, most tsunamis originate thousands of miles away, usually off Alaska, Kamchatka, or South America, so there is a relatively long interval between the earthquake tremors on the seismo-

Fig. 4-1. The second wave of the April 1, 1946 tsunami approaching Kawelo Bay on the north coast of Oahu. Shortly after the photo was taken, water swept over the low ridge where the writer had been standing.

graph and the arrival of the waves. Earthquakes tremors travel through the earth at about 5 miles (8 km) per second, and the water waves, set up by the faulting of the bottom, move across the deep open ocean at about 450 miles (725 km) per hour. Therefore, it is possible to warn the Hawaiians by rapidly locating the giant earthquakes, using records from seismographic stations, and then (knowing their speed of travel) estimating the time when the sea waves will reach the islands. Sirens all along the Hawaiian coasts are set off with lusty blasts, making it possible for people to evacuate the low ground. Warnings never come from ships at sea, because the tsunami waves are so long, low, and gently sloping over deep water as to be imperceptible. It is only in shallow water near the coast that they may rise to great heights.

Hawaiian Experience in 1946. On April 1, 1946, my wife and I were enjoying our first post-World War II vacation, living in a little house on a low beach ridge along the north coast of Oahu. We were awakened at 6:30 in the morning by a great hissing sound, just as if there were hundreds of locomotives blowing off steam. Rushing to the front window we saw the sea swirling up around the front of our cottage. As we watched with amazement, we saw the water start to recede. Down it went, exposing first the beach and then the outer coral reef, and finally the small lagoon inside the reef, showing the fish jumping on the exposed sea floor. As a scientist, my first thought was, "Let's get a photo of what is happening." I pulled on a jacket and sneakers and ran out with my camera. My second thought was, "Is this a tsunami? How could it be? They haven't had any large tsunamis since 1877." Standing on the beach ridge, I was soon disillusioned by seeing the water start to break violently over the nearby reef. As it advanced, I stood there taking photographs and thinking, "To bad, I didn't get a shot of the earlier wave, this one surely won't rise so high." Pretty soon the water began to surge above the horizon and I got really worried (Fig. 4-1). Just in time, I ran behind our house where my more prudent wife had already stationed herself. The advancing wave smashed the glassed-in porch at the front of

Fig. 4-2. The completely destroyed house next to that in which the writer was living at the time of the tsunami.

the cottage, and we saw the refrigerator being carried past us into the low-lying sugar cane fields. Looking along the ridge, we were stunned to see that the cottage next to us had been completely destroyed (Fig. 4-2) while we stood safely behind our partially wrecked dwelling.

Out went the second wave, and we started running along the beach ridge to where we knew there was a path through the cane field leading to the elevated road, our best possible escape. We found a camp with some sodden

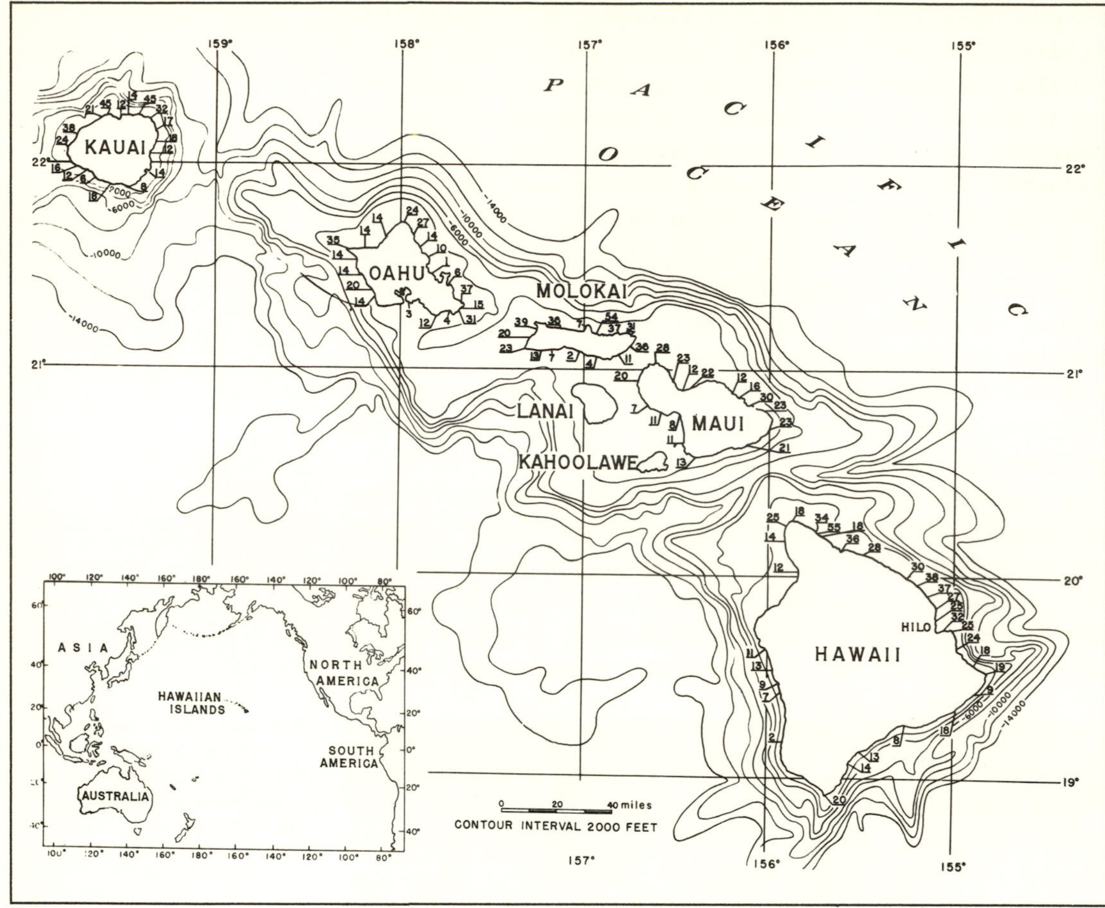

Fig. 4-3. The heights in feet to which the water rose during the 1946 tsunami along the coasts of the major Hawaiian Islands. Contours show the relief of the sea floor adjacent to the islands.

and terror-stricken Hawaiian women on the low ridge, and persuaded them to join us in our escape route. As we ran across the low cane field, we heard another wave following us and crushing the cane behind us. We all arrived just in time to climb onto the road, where we found other refugees.

After watching five more waves come in at about 15-minute intervals, I decided that the waves surely must have now become diminished in height so I could go back and start rescuing some of our things. Again I was wrong. The eighth wave was the highest of all, and that dry spot behind our partly demolished house was now completely inundated. I rushed to a nearby ironwood tree and climbed above the flood, hanging on as the tree swayed back and forth in the advancing wave. When it was over, I forgot about our belongings and returned to the road, still in pajamas and carrying a camera, now with film exhausted.

Later, I telephoned two geologists who were glad to join me in making an island study of the effects of the waves (Shepard et al., 1950). What we had experienced was nothing compared to the 30-foot (9-m) waves that flooded much of Hilo and other Hawaiian cities, killing 150 people. In fact, the water had risen as high as 56 feet (17 m) in one valley we visited along the coast of Hawaii. Heights of waves we measured are shown in Fig. 4-3. Among the conclusions from our study were:

- The first tsunami wave may not be the highest, a conclusion which seems to have been overlooked in previous literature.
- The highest waves in the Hawaiian Islands are on the side of an island from which the waves are advancing, but some waves go around a corner without much loss of height.
- Tsunamis may attack an apparently protected coast by reflection off a continental margin, just like a carom shot in billiards.
- Coral reefs are effective dampeners of tsunamis; the waves were negligible inside the broad reef at Kaneohe Bay.

Other tsunamis. Of the 50 American states, Hawaii is the most prone to tsunami attack, in spite of the fact that it is located thousands of miles from the great earthquake belt along the Pacific margins. Tsunamis can cross wide oceans with a relatively small loss of energy. Whereas there had been no large tsunamis in the islands between 1877 and 1946, three more of these destructive waves had battered the Hawaiian shores by 1960. These came, respectively, from the sea floor off Kamchatka (1952), from the Aleutian Trench (1957), and from Chile (1960).

The West Coast of the United States has been very free of the large waves that hit Hawaii, despite the occurrence of many earthquakes off the coast of California. It happens that the fault movements off California that produce earthquakes have mostly horizontal shifts, like the road and fence displacements that accompanied the San Francisco earthquake in 1906. Such movements do not lift or drop the ocean bottom, which is the usual cause of the tsunamis. In 1964, for the first time in history, waves that hit Northern California and parts of Oregon caused serious damage to Crescent City and other towns. At this time, southwestern Alaska experienced one of the world's greatest earthquakes. Because of all the coastal inlets along the Alaskan coast, local waves were not very destructive, although they did wreck portions of Valdez—the southern terminus of the Alaska pipeline. However, at this time, for some unknown reason, the large waves moved down the West Coast of America, in contrast to those that accompanied previous Alaskan earthquakes (Plafker and Kachadoorian, 1966). Crescent City was in their path and received the brunt of the attack. Surprisingly, the Hawaiian Islands had almost no wave damage on this occasion. Tsunami siren warnings wailed in vain.

In 1755, Europe experienced its worst tsunami in recorded history. Faulting along the Mid-Atlantic Ridge near the Azores caused waves to sweep onto the Portuguese coast and follow up the Tagus Estuary, swamping the narrow, heavily populated coastal plain at Lisbon.

The greatest tsunami waves of which there are good records occurred in 1883, when the volcanic island of Krakatoa, between Java and Sumatra, underwent volcanic explosions and a large engulfment. The latter sent waves around the world and were actually observed in the English Channel. At the nearby Indonesian islands, waves rose more than 100 feet (30 m) and drowned tens of thousands of people. It seems likely that a similar occurrence virtually wiped out the Minoan civilization about 1470 B.C. The volcanic island, Santorini, was largely destroyed at this time, and huge waves must have been set up in the Aegean Sea, swamping the lowlands of Crete and many other areas (Bascom, 1976). Curiously, the legend of the Lost Continent of Atlantis seems to have originated from this disaster, and centuries later the story diverted the disaster to a supposed ancient island of the Atlantic, as told by the great Greek philosopher, Plato.

Storm Surges

By far the worst inroads of the ocean onto the coasts of the United States and many other countries, including India and Bangladesh, have resulted from hurricanes and the equivalent tropical storms, called *typhoons*. As a storm center moves toward the coast it is usually accompanied by a rise in sea level. The water level may come up as much as 50 feet (15 m), covering many miles of coastal lowlands. Thus in 1900, a hurricane came into the Texas coast at the city of Galveston, and the sea-level rise destroyed the low sea walls and engulfed the whole populated area, drowning 600 people. The sea only rose about 15 feet (4.5 m) at Galveston; whereas in a subsequent storm along the Gulf Coast, the 1969 hurricane, called Camille, rose as much as 30 feet (9.1 m) along the Alabama coast (Fig. 4-4), but caused far less loss of life due to hurricane warnings and motor transportation, allowing an easier method of escape than at Galveston. Other hurricanes have caused great destruction along the East Coast, notably the 1938 New England hurricane, where 548 people were lost.

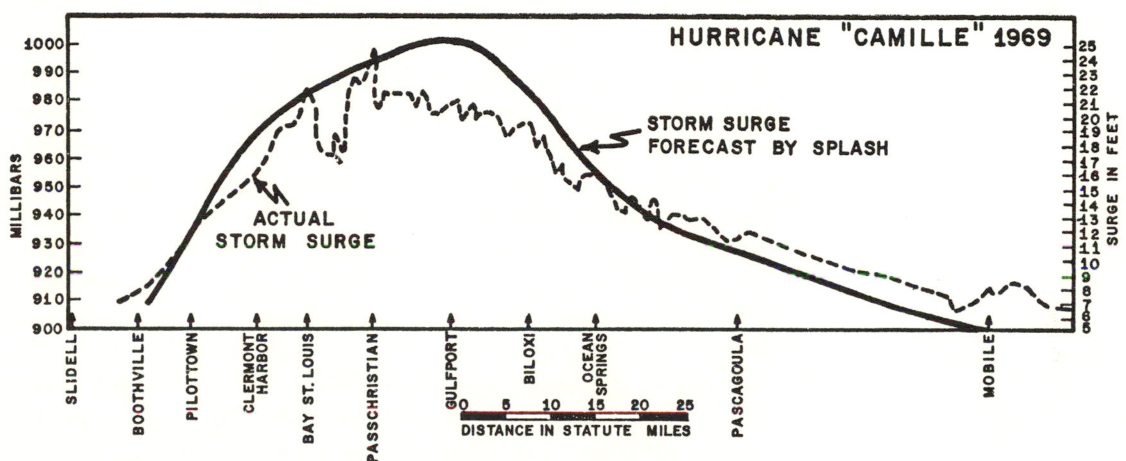

Fig. 4-4. The height of the storm surge in Hurricane Camille along the Gulf Coast. Courtesy of the National Oceanic and Atmospheric Administration, Hurricane Center.

Our disasters from storm waves are trivial compared with those at the head of the Bay of Bengal. Perhaps the greatest loss of life in a natural catastrophe of all time took place in 1737 in the country now called Bangladesh. A typhoon moved up the Bay of Bengal, funnelling water ahead of it and swept over the low plains of the Ganges-Brahmaputra Delta, drowning 300,000. China recently announced that 660,000 died in its 1976 earthquake.

Storm surges differ from tsunamis in that they do not have the alternating rise and fall of water level that continues for hours or even days. The water comes in in advance of hurricane centers, rising somewhat like the tide, but usually the rise is much accentuated by the huge wind waves. Because the water rises much slower than the waves of a tsunami, it is often possible to flee to high ground ahead of the advancing flood. Unfortunately, the violence of the storm and the devastating rains make escape very difficult. Better warning systems are important. When the weather bureau makes good forecasts, people have ample time to get out of the threatened area, but the hurricanes sometimes change their course a short time before coming ashore.

The effects of both hurricanes and tsunamis on the coastline are spectacular, as will be discussed in a subsequent chapter.

Landslide Surges

A wave from a landslide is quite distinctive from the other types. It can be compared to what happens when a child gets in at one end of a bathtub and slides into the water, pushing water over the other end and causing a family crisis. In July 1958, the largest wave in human record was produced in this way near the head of Lituya Bay, an Alaskan fiord (often spelled fjord). An earthquake caused a huge mass of rock to break from a mountain and fall precipitously into the deep fjord. The displaced water crossed to the other side of the inlet and actually rose up the wooded slope to 1,700 feet (518 m). I was in the Washington office of the Coast and Geodetic Survey shortly after this event and heard some of their wave experts saying, "This is impossible. Nothing like this has ever happened." However, the late Don Miller of the U.S. Geological Survey made a careful study (1960) and found the trees all tipped over upslope by the force of the wave at this amazing height (Fig. 4-5).

Fortunately, no one was living along the shores, but three fishing boats were anchored in the lower end of the bay. How any of the fishermen survived is a mystery, but we do know that the wave that swept down the bay was much lower than 1,700 feet (518 m) high. The seaward continuation of the wave cut a swath into the slopes on either side at about 100 feet (30 m). One boatman remained anchored but by steaming into the approaching wave was able to ride up over the crest; it was partly because the anchor held sufficiently before the cable broke, so that his boat was not swept out of the bay. A second boat disappeared with no sign of survivors, but the most amazing adventure was experienced by the occupants of the third boat. They were lifted by the wave and carried over the spit at the bay mouth so that they could actually look down on the trees as they passed by and over them. Landing in the open ocean outside, their boat sank but they had time to get into their skiff and save their lives.

Fig. 4-5. The swash developing from the large rockfall into Lituya Bay in 1958. The arrow shows where the wave cut away the trees up to 1700 feet (518 m) above sea level. The front of Lituya Glacier (right) was cut back 1800 feet (549 m) by the slide. From D. J. Miller, 1960. Printed by U.S. Gov. Printing Off., Wash., D.C.

These landslide surges differ markedly from a tsunami. Apparently there was only one large wave and, despite its 100-foot (30-m) height, as it came down the lower fiord it dissipated rapidly after crossing the spit, so that no boatmen recorded it on the outside. In the bay the front of the wave was not steep enough to become a breaker, otherwise the two boats that rode the wave would have capsized.

Landslides have had more tragic effects, as in the Inland Sea of Japan, where, in 1792, in Shimbara Bay of Kyushu Island, a fall at the time of an earthquake produced three large waves that drowned 15,000 people along the edge of the bay. Fortunately, most great avalanches do not carry their debris into bodies of water where there are low lying towns to be engulfed by the waves. Damage from this source is far less than from tsunamis and storm surges.

References

Bascom, W., 1976. "Science and ancient sea stories." *Oceans,* v. 9, no. 4, pp. 10-11.

Miller, D. J., 1960. "Giant waves in Lituya Bay, Alaska." *U.S. Geol. Surv. Prof. Paper 354-C,* pp. 51-86.

Plafker, F., and R. Kachadoorian, 1966. "Geologic effects of the March, 1964 earthquake and associated seismic sea waves on Kodiak and nearby islands, Alaska." *U.S. Geol. Surv. Prof. Paper 543-D,* pp. D1-D46.

Shepard, F. P., G. A. MacDonald, and D. C. Cox, 1950. "The tsunami of April 1, 1946." *Bull. Scripps Inst. Oceanog.*, Univ. Calif. Press, v. 5, no. 6, pp. 391-527.

Suggested Supplementary Reading

Adams, W. M., ed., 1970. *"Tsunamis in the Pacific Ocean." Proc. Intl. Symp. on Tsunamis and Tsunami Research.* East-West Center Press, Honolulu, 513 pp.

Bascom, W., 1964. *Waves and Beaches: the Dynamics of the Ocean Surface.* Anchor Books, Doubleday & Co., Inc., Garden City, N.Y. Ch. 6.

Van Dorn, W.G., 1974. *Oceanography and Seamanship.* Dodd, Mead & Co., N.Y. Ch. 18.

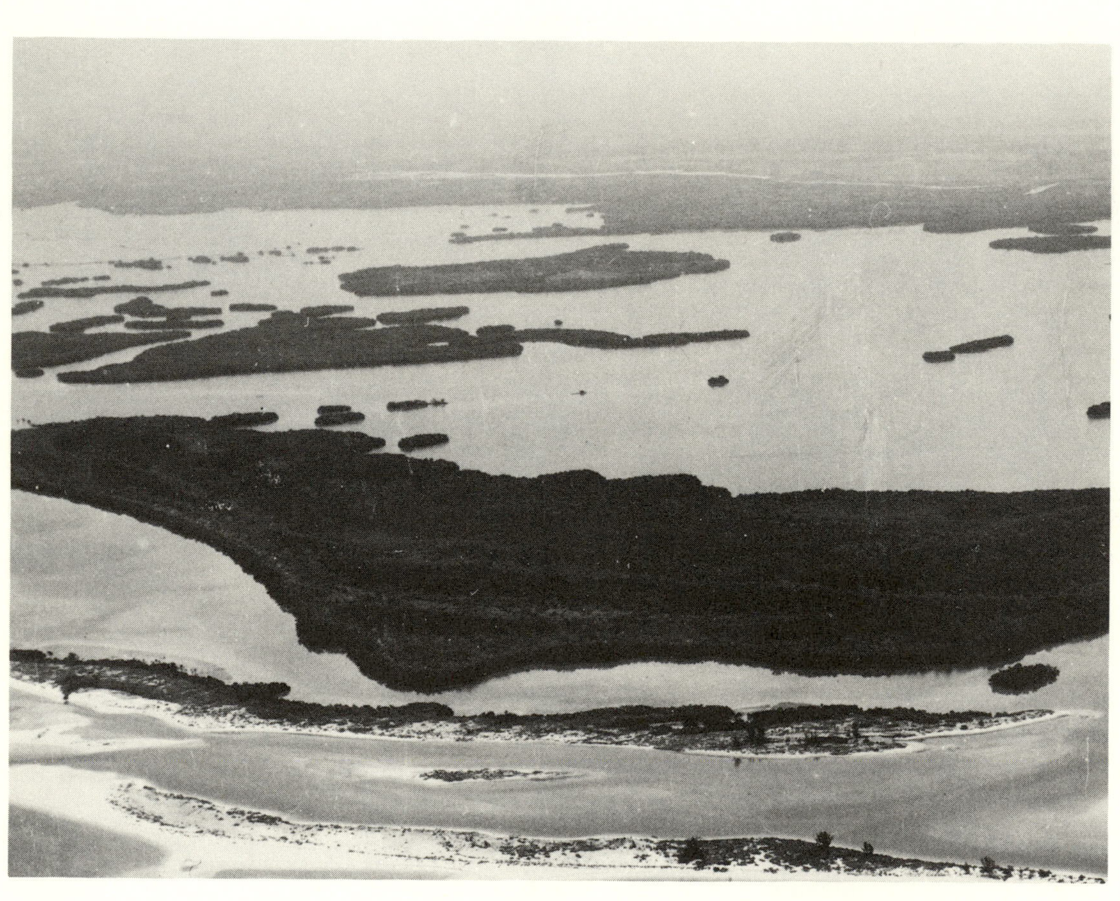

Why are coasts so different in character?

Chapter 5

If you look at a map of the East Coast of the United States you will see a great contrast in the configuration of the shoreline. To the north is the deeply indented Maine coast with all its hilly islands; off Cape Cod, a straight-cliffed coast followed by a low but smooth shoreline; the straight, outer coast of Long Island and New Jersey with long sandy islands bordered landward by lagoons and marshes; the deeply indented Delaware and Chesapeake Bay with almost no islands; then the sand islands with cuspate points, including Capes Hatteras, Lookout, and Fear that border North Carolina; then more indentations and straight sand islands off South Carolina and Georgia; and, finally, the long straight beaches of Florida with coral keys at the southern tip. How do we explain these contrasts and the many other types of coast of the world?

When I first taught beginning geology, instructors used to show maps of embayed coasts and explain them as due to the sinking of the land, allowing the ocean to come up into the river valleys. These we called *coasts of submergence.* The straight lowland coasts were explained by uplift of a supposedly even, sloping sea floor that had been beveled by the waves before uplift. We called these *coasts of emergence*, and included those bordered by long sandy islands and raised terraces. The third principal type had deltas built out into the sea by rivers, like the Mississippi Delta. These we called *neutral coasts* because it was thought that they had neither risen nor sunk. *Compound coasts* supposedly had undergone both sinking and rising so that they had indentations inside and straight sand islands on the outside. In addition, special types of coast included *fault coasts, volcanic coasts,* and *coasts built by organisms.*

This classification was made famous by the Columbia University professor Douglas Johnson (1919) and his hundreds of devoted students. But in 1937, as a young man, I took it upon myself to criticize this simple classification and immediately found myself in a hornets' nest. However, I was not happy teaching my students that those indented coasts were explained by submergence when, as in the case of Maine and Norway, they are well known to have been rising because of weight loss due to the retreat of the thick ice caps of the glacial epoch, which started to retreat approximately 18,000 years ago. Also, many of the straight coasts had narrow sand barriers bordering lagoons with drowned valleys inside (Figs. 5-1, and 5-2[8]). Where wells had been drilled into the long sand barriers, the well logs usually showed evidence that the

Fig. 5-1. A drowned valley coast straightened by a barrier on the outside of Martha's Vineyard Island, Massachusetts. Originally, this was an outwash plain from a glacier, later cut into by streams; subsequently it was drowned prior to the building of the barrier. Aerial photo from U.S. Coast and Geodetic Survey.

coasts had been sinking, rather than rising. As for the neutral coasts, the geological literature showed unmistakable evidence of submergence. Even the old lighthouses built almost 100 years ago at the edge of the Mississippi Delta appear on more recent maps as "abandoned" and surrounded by water. But the most fundamental objection to the Douglas Johnson classification was, and still is, that the melting of the great ice sheets, which occurred mostly between 18,000 and 3,000 years ago, was accompanied by a rise in sea level of some 400 feet (122 m). Since most coasts had been relatively stable during this period of time, the rising sea level must have drowned the lower ends of the river valleys over most of the world. To add to the difficulties of the classification of a coast as one of emergence, the criterion was often used that wave-cut beaches and sea cliffs are found well above sea level, indicating a coast of emergence. However, as we now know, there were times during the two million years of the glacial epoch when the ice caps were smaller than they are today, and therefore the sea stood higher along all stable coastal areas. Hence, the supposed emergence might be due to remnant wave-cut benches formed when the sea stood high, and therefore this was a world-wide phenomenon.

Replacing the Old Coastal Classification

For these and other reasons that are too numerous to include here, it seemed necessary to attempt a new coastal classification that would give appropriate consideration to sea-level changes, which are universally accepted by geologists as having accompanied the waxing and waning of the great ice sheets. Of course, some coasts have sunk and others have been pushed up, but unless one has clear evidence of such actual movements (and it is usually not evident, especially on maps or in photographs) one has to use with care any evidence, such as drowned river valleys or elevated terraces, before calling a coast one of submergence or emergence. Any truly neutral coast that has sufficient relief so that rivers could cut valleys into it during glacial stages of low sea level will have the indentations that are called *drowned valleys* (Fig. 5-2[1]) unless

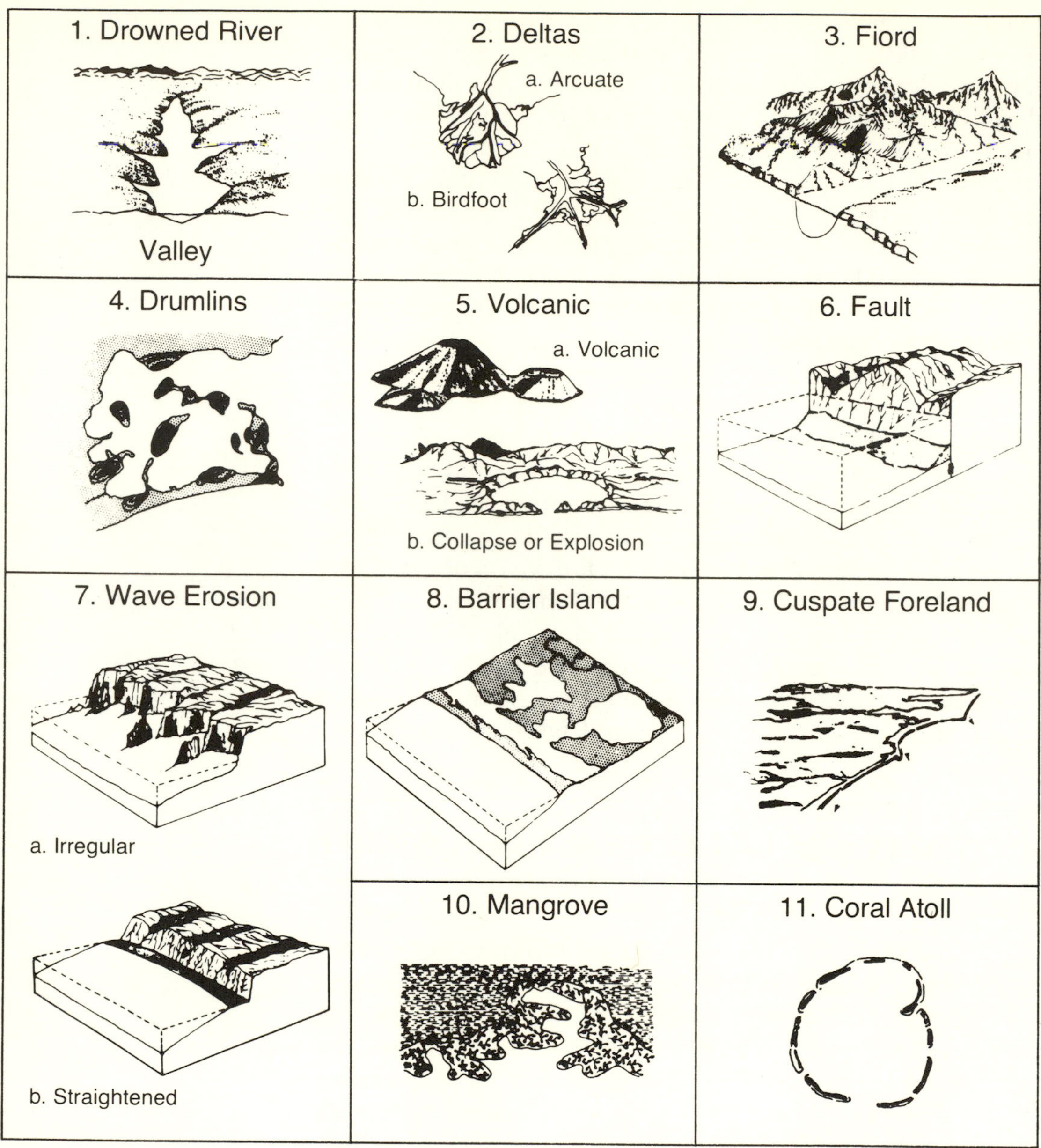

Fig. 5-2. Some types of coasts included in the classification. From F. P. Shepard, *Submarine Geology,* 3rd edition (New York, Harper & Row, 1973).

these have been filled since the rise in sea level virtually stopped a few thousand years ago.

If a coast is essentially in the same condition in which it was left at the end of the last sea-level rise, why not call it a *primary coast;* whereas if it has been changed appreciably by the waves or other processes after sea level stabilized, let's call it a *secondary coast.*

Primary Coasts

Land Erosion Coasts. If the shape of a coast has a pattern like the outline of an oak leaf (Fig. 5-2[1]), it is probably the result of the drowning of the coastal river valleys, and we call it a *ria coast* from the Spanish word for river. Another type of deeply indented coast is due to glacial erosion of old river valleys. This is called a *fiord coast* from the Norwegian embayments. Fiords (also spelled fjords) can usually be recognized by their having relatively straight sides, in contrast to the usual curving sides of the ria shorelines and by the abundance of rock islands. Also, looking at a coastal chart one will almost always find water in a fiord that is hundreds or even thousands of feet deep, and usually deeper well up the fiords than in the lower end (Fig. 5-2[3]). The ria-type coast usually has shallow water rarely more than 100 feet (30 m) deep and generally shoaling farther up the embayment. Glacial erosion also produces many rocky islands, allowing yachts and cruise ships to traverse beautiful inside passageways, like the coasts of British Columbia and southeastern Alaska.

Coasts Built Out by Land Processes. In many places, sediments brought from the land have built out the coasts so actively that the shoreline is now quite unrelated to the postglacial sea-level rise. Most notable among these types are the deltaic coasts. Many of the great rivers of the world have extended the coast at their mouths into large convexities. The fourth letter in the Greek alphabet, Δ (delta), originated because of the triangular shape of the Nile Delta. Deltaic coasts have other shapes, the three main types include lobate, digitate (Fig. 5-2[2]), and cuspate (Fig. 5-3). The great Mississippi Delta includes both lobate and digitate. The latter, called appropriately the Birdfoot Delta, is known to have been built during the past 2,000 years (Fisk et al., 1954). Because waves along open coasts are so effective in carrying away the sediments brought in by streams and because of the recent rise in sea level that drowned many deltas, it takes a large river to build a delta along the open coast, and most of them are located where there is a broad indentation of the coast, as at the head of the Bay of Bengal, where the Ganges-Brahmaputra Delta

Fig. 5-3. A large rock slide at Humbug Mountain near Port Orford, Oregon. Note the extension of the coast due to the slide. From University of California, Hydraulic Engineering Laboratory, 1948.

has a front 200 miles (320 km) long, and the great Niger Delta of West Africa. Even the Mississippi Delta filled an indentation in the Gulf Coast that extended up as far as Cairo, Illinois, a few million years ago.

The continental glaciers of the Ice Age carried large quantities of sediments and rocks of all sizes to their outer margins and, in their retreat, left behind them hills and ridges (called *moraines*), some of which still stand above the seas that advanced as the glaciers retreated. These coasts we call *glacial deposition.* For example, Long Island is largely a glacial moraine straightened by the sea, and the oval-shaped hills in Boston Harbor are a type of glacial dump, which we call *drumlins* (Fig. 5-2[4]). Along the sides of Puget Sound at Seattle, the coast is partly a glacial moraine, although it has been considerably shaped by glaciers moving along the inlets that characterize the coast.

Locally, landslides carry great masses of debris into the ocean, building out the coast. An example of a *landslide coast* is along the Oregon shoreline (Fig. 5-3). However, the great slide in Lituya Bay that caused a 1,700-foot (518-m) swash (Fig. 4-5) did not extend the coast because of the deep water and the steep side of the fiord where the rock mass fell. A very common type of landslide coast can be recognized by finding an irregularly topped terrace along the base of a sea cliff (Fig. 5-4). These had sometimes been mistaken for elevated wave-cut terraces. Generally, the base of such a landslide terrace shows a mass of boulders extending out into the ocean.

Volcanic Coasts. Most islands that rise above the deep ocean are volcanic. If the volcanism has been of recent origin, the coasts consist of the latest lava flows (Fig. 5-2[5a]). Like a delta, these coasts have great lobes where the lava has built out into the ocean. Usually the slope of the adjacent sea floor is approximately the same as the

Fig. 5-4. The irregular topped terrace formed by a landslide north of La Jolla, California. The rocks at the toe of the slide are a concentration of boulders caused by wave erosion.

lava flow at the flank of the volcano on land. Volcanic coasts are common around the south and west sides of the island of Hawaii, where many flows have occurred since the United States' annexation of the islands in 1899. Another type of volcanic coast is due to the collapse of one side of a volcano where lava has flowed down from underneath and escaped through a crack in the submerged side of the volcano. This produces a concave, instead of the usual convex, shoreline of a lava flow (Fig. 5-2[5b]). If you have swum at Hanauma Bay, east of Honolulu (Fig. 5-5), you have seen one of these partially collapsed volcanoes. The head of the bay has a small coral reef of more recent origin.

Coasts Due to Earth Movements. Where the earth's crust is being warped or faulted, what we call *diastrophic* coasts may develop. This type can be easily recognized where an escarpment at the coast continues as a steep slope beneath the sea (Fig. 5-2[6]). If this escarpment were simply a sea cliff, it would be bordered at the coast by a gently sloping platform (Fig. 5-2[7b]). However, if the seaward side of a fault has been moving down at an appreciable rate, it will lack the bench that is due to the wearing back of a cliff by wave erosion. As an example of a fault coast, if you are cruising along the straight northeast side of San Clemente Island off Southern California, you will find deep water near the steep straight shore. Even more striking is the part of the straight west coast of the Gulf of California where, within a half mile (0.8 km) of the shore, the water is consistently several thousand feet deep.

Fig. 5-5. Volcanic craters east of Honolulu, Oahu. The collapse of one side of the crater (right) has produced Hanauma Bay.

Diastrophic coasts cannot always be recognized: if, for example, the land has been downwarped, producing drowned valleys, this might be equally well due to the postglacial rise in sea level. On the other hand, a fault coast may have been uplifted or sunk at the time of a great earthquake, as happened in many places along the Alaskan coast in 1964. On one side of Prince William Sound, a wave-cut terrace was brought up above sea level at the time of the earthquake, exposing a quarter mile (0.4 km) of the old sea-cut bench (Fig. 5-6).

Fig. 5-6. A diastrophic coast in Prince William Sound, Alaska, where a wave-cut terrace was raised above sea level during the 1964 earthquake. Photo by George Plafker, U.S. Geological Survey.

Secondary Coasts

Wave-Erosion Coasts. Where waves have attacked the coast sufficiently since the sea-level rise so that their effect is more pronounced than the drowning of the valleys or the submerging of any other feature, we have a secondary coast attributable to wave erosion (Fig. 5-2[7]). The waves may have two quite different effects. If the coast they have attacked is made of material of relatively even consistency, the shore is generally straightened, and the shore cliffs are bordered by a gently sloping bench and sea floor (Fig. 5-2[7b]). However, if the waves attack a coast of variable hardness, the land will have a very irregular shoreline (Fig. 5-2[7a]; Plate 2, Fig. 4). This type of coast somewhat resembles the drowned valley ria coast, but with only minor indentations. It lacks long embayments, such as Chesapeake Bay.

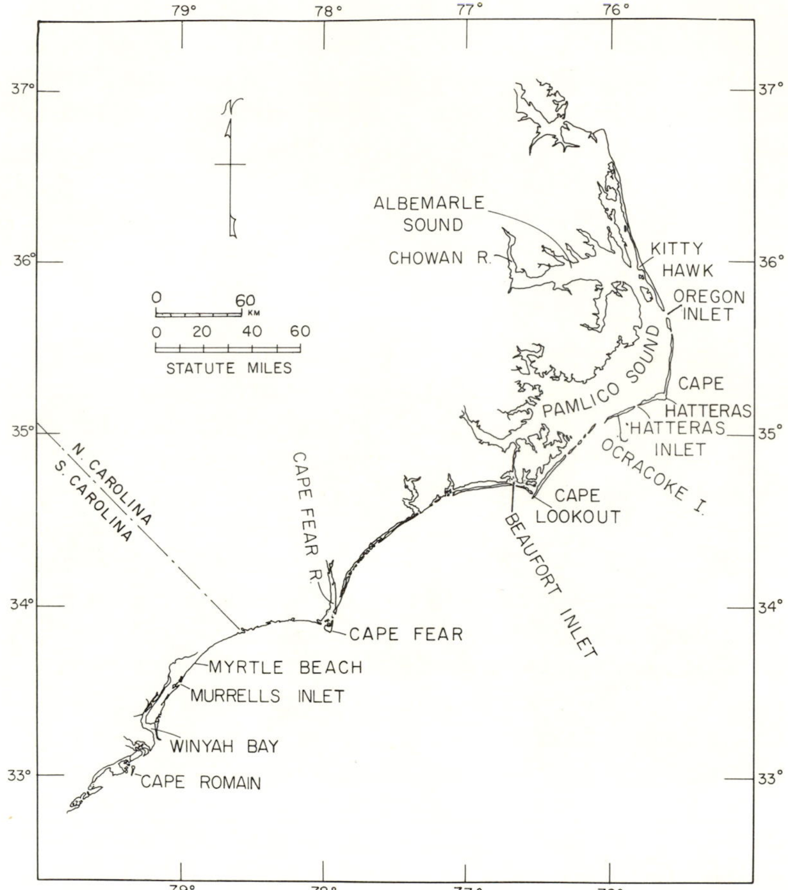

Fig. 5-7. The cuspate forelands extending from Cape Hatteras to Cape Romain. These are the finest examples in the world and are related to back eddies from the Gulf Stream.

Marine Deposition Coasts. Since the postglacial rise in sea level, the sea has taken over the shoreline characteristics quite effectively. Where the waves break well out from shore due to gentle slopes, sand is pushed up, forming sandbars, and these in turn may grow into long sand islands, called *barrier islands* (Fig. 5-2[8]; Plate 2, Figs. 1, 2). These islands may be widened by the addition of sand to their seaward edges, until they are miles across. Many coastal towns of the southern and eastern United States are built on barrier islands—for example, Atlantic City, Miami Beach, and Galveston. Barrier islands are built up partly by the winds that develop sand dunes. The barriers are often connected to the mainland, at least at one end, but they usually have a shallow-water lagoon on the inside. Much of the Intracoastal Waterway along the east and south coasts of the United States is the result of dredging of lagoons inside barrier islands. It has been estimated that barriers constitute 47 percent of the coast of our 48 contiguous states. Barriers are common elsewhere, as in the Netherlands, but they are most common in the United States. Barrier islands often develop along deltaic coasts where a portion of the delta is inactive, as at Grand Isle, Louisiana, or at Alexandria, Egypt.

A special type of barrier island is referred to as a *cuspate foreland* (Fig. 5-2[9]), of which Cape Hatteras (Fig. 5-7) is the prime example. These symmetrical sand points off the Carolinas are the result of giant eddies inside the Gulf Stream. However, a similar set of cuspate forelands off the northwestern coast of Alaska has not clearly been explained in this way. Little is known about the currents in that area and the coast is icebound most of the year. The United States has almost a monopoly on cuspate forelands. Cape San Blas, in northwestern Florida, is another good example, and smaller ones are located farther south. Why are they virtually missing in the rest of the world? We simply do not know the explanation. Cape Verde in Senegal and Monte Argentario, north of Rome, are cuspate in outline but are actually land-tied rock islands, called *tombolos,* like Pt. Sur (Fig. 5-8) along the California coast.

Another common type of coast, which is at least in part due to marine deposition, is the mud flat and coastal marsh, such as exist along the Everglades of southwest Florida. Wherever the post-glacial sea-level rise stopped on a very flat, slightly submerged plain, an occasional high sea level may carry in enough sediment to produce a marshy shoreline coast that is usually submerged during times of high tides.

Coasts Built Forward by Animals and Plants. In tropical areas, coral reefs abound. The corals cannot build above low tide, but, having produced a slightly submerged reef along the coast, the waves during great storms can and do build up the reef margins above sea level. Also, some coral-reef coasts, such as the Florida Keys, were formed during an interglacial stage of the Ice Age when the sea stood about 20 feet (6 m) higher than at present. Most of the coral-reef islands of the Pacific, such as the Carolines, the Marshalls, and the Gilberts, are so low that a great storm, like the one described in *The Hurricane* by Nordhoff and Hall (1936), will virtually submerge the entire land mass. Therefore, it seems likely that these islands are the result of great storms tearing off large blocks from the reef margin and building up the reef on the inside.

Other coasts have been formed by mangroves that can grow as trees in salt water. A barbed seed pod falls from a mangrove tree overhanging the shoreline and starts a new plant in the bottom, or the seed might drift out to a shoal and get rooted there. In this way many islands and projections of the mainland develop along coasts, such as in southwestern Florida (Fig. 5-9), where they have greatly extended the mainland, partly by currents depositing sediment around the mangrove roots and filling in the bays. Other shallow-water areas may be converted into land by the growth of the tall salt-water marsh grass, such as *Spartina*. This also entraps sediment, developing new land areas.

Fig. 5-8. Pt. Sur, south of Carmel, California, an island connected to the mainland by a sandspit (tombolo). Photo by D. L. Inman.

Fig. 5-9. Mangrove islands in Estero Bay near Fort Meyers, west coast of Florida. The islands in the lower part of the picture are barriers developed in recent years. Photo by D. L. Inman.

Discussion and Possible Shortcomings of the Classification

The preceding gives the gist of a coastal classification that I have been building up (and I hope improving) during several decades. It certainly could be revised further. Much more complicated classifications have been suggested. Such names are proposed as: convergent and divergent coasts, low and high energy coasts, stable and mobile region coasts, Atlantic and Pacific coasts, collision coasts, and marginal coasts. Many of these terms have scientific importance; others are hard to comprehend. However, the chief reason for not including them here is that I do not think one could make use of the terms in looking at a coast, whether from the air or from the ground, or from studying a map or nautical chart. The classification that I have provided, along with the illustrations, should allow you to decide how most of the coasts have been formed without consulting a treatise written in technical terms by a geologist.

An important point to remember in classifying coasts is that many of them come under two or more categories (Pl. 2, Fig. 3; Fig. 5-9). Many other examples could be given. Certainly there is no objection to giving a coast a multiple classification.

References

Fisk, H. N., E. McFarlan, Jr., C. R. Kolb, and L. J. Wilbert, Jr., 1954. "Sedimentary framework of the modern Mississippi Delta." *Jour. Sed. Petrology,* v. 24, no. 2, pp. 76-99.

Johnson, D. W., 1919. *Shore Processes and Shoreline Development.* Wiley, N.Y., 584 pp.

Nordhoff, C., and J. Hall, 1936. *The Hurricane.* The Atlantic Monthly Press, 257 pp.

Shepard, F. P., 1973. *Submarine Geology.* 3rd Ed., Ch. 6. Harper & Row, N.Y., 517 pp.

Suggested Supplementary Reading

King, C. A. M., 1972. *Beaches and Coasts.* 2nd Ed., St. Martin's Press, N.Y. Ch. 5.

Shepard, F. P., and H. R. Wanless, 1971. *Our Changing Coastlines.* McGraw-Hill Co., N.Y. Ch. 16.

Will our beaches disappear?

Chapter 6

In a voluminous report on the *Beaches of California,* the California Coastal Commission makes dire predictions on the future of Southern California beaches. Simply stated, the Commission maintains reasonably that the chief source of sand for these beaches is wet-weather streams that come down from the mountains during rainy periods. They state that most of the stream valleys have now been dammed and the sand collects in the artificial lakes behind these dams. Does this mean that the millions of happy people who throng the beaches of Southern California each summer are soon to be deprived of this pleasure? I find it difficult to become too gloomy. For forty years I have been following with interest the waxing and waning of Southern California beaches, and must say that to date the principal change has been in the bathing suits. Most temporary losses have been corrected by man. The long beach on the north side of La Jolla, California, should have been disappearing. Sources of sand have been largely cut off, yet the beach is actually growing wider. Prior to 1946, the winter storms cut away almost all of the sand for a quarter of a mile (0.4 km), leaving cobbles and exposing older formations (Fig. 6-1A). But the sand returned each summer (Fig. 6-1B). Since 1964 we have had virtually continuous broad sand beaches in this area throughout the year. Thus we encounter an enigma, and something seems to be missing from our simple explanation, as in the case of the emerging and submerging coasts of the Johnson classification discussed in the previous chapter.

Fig. 6-1. Showing the (A) winter and (B) summer conditions of the beach south of Scripps Institution at La Jolla, California. The cutting away of the beach in winter allowed the waves to attack the alluvial cliffs (note blocks in (A) from recent slump). In 1947, conditions changed so that the sands persisted and retreat of the cliffs virtually stopped. See also Fig. 7-7.

Beach Nomenclature

Figure 6-2 starts us off with a profile of a typical beach. We can assume that the beach terminates landward at a sea cliff, a row of sand dunes, or at a sea wall. Often the inner portion of the beach is relatively horizontal. This is called a *berm*, and sometimes there are several berms at increasing heights toward the inner margin. We call this part of the beach the *backshore*. Farther seaward, we encounter a slope, and this is referred to as the *foreshore*. If it flattens near low tide level, this is called the *low tide terrace*. Usually, seaward of the foreshore there is a depression, called

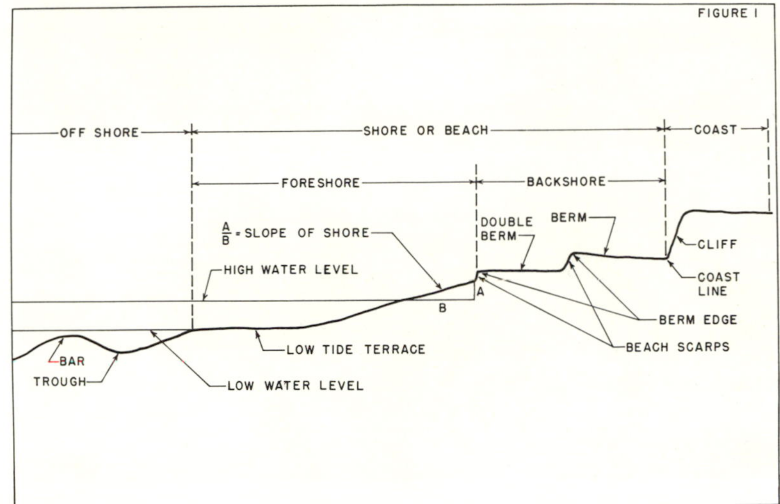

Fig. 6-2. The principal subdivisions of typical beaches and of the adjacent shallow-water area.

a *longshore trough*, which may be several feet deep if it contains the feeder of a rip current. Outside the trough there is usually a sand ridge, called a *longshore bar*. In some gently sloping beaches, a series of bars and troughs run parallel to the shore. When flying relatively low along the shore, these are clearly seen by the lines of breakers along much of the Gulf Coast of the United States (Fig. 6.3).

Beach Contrasts Between Storm and Low-wave Conditions

If we compare beach profiles made after storm waves or large swells have attacked the shore with earlier profiles made after periods of small waves (Fig. 6-4), we find the storm berm greatly decreasd in width or even totally eliminated. Occasionally an almost vertical sand cliff may exist at the outer edge of the berm (Fig. 6-5), but this will only last for a short time because the sand, after drying,

Fig. 6-3. Alternating bars and troughs along the gently sloping shore of the Gulf Coast, indicated by the several lines of breakers. Photo taken by D. L. Inman off Padre Island, a barrier island of southern Texas.

will collapse. The low-tide terrace is also eliminated by many storms so that the foreshore slopes out as far as the longshore trough, the latter somewhat deepened. Also, the longshore bar has moved seaward and has greater relief, although its crest is now much deeper below sea level.

As an example, at La Jolla, California, where the waves converge on a rocky point, a sandy beach (called Boomer because of the pounding of the breakers and the interesting but dangerous surf riding) exists during the summer. The sand may disappear in a few hours when the first large waves of fall approach it from a northwesterly direction, leaving a mass of cobbles (Fig. 6-6A and B). Actually much of the sand is carried south along the shore, and a somewhat smaller beach forms at the southerly end of the indentation. During the following spring or early summer, Boomer Beach is gradually rebuilt, because now the swells begin to approach from the south, coming as they do from those great storms of the Southern Hemisphere that were discussed in Chapter 3. This sequence is typical of many other areas.

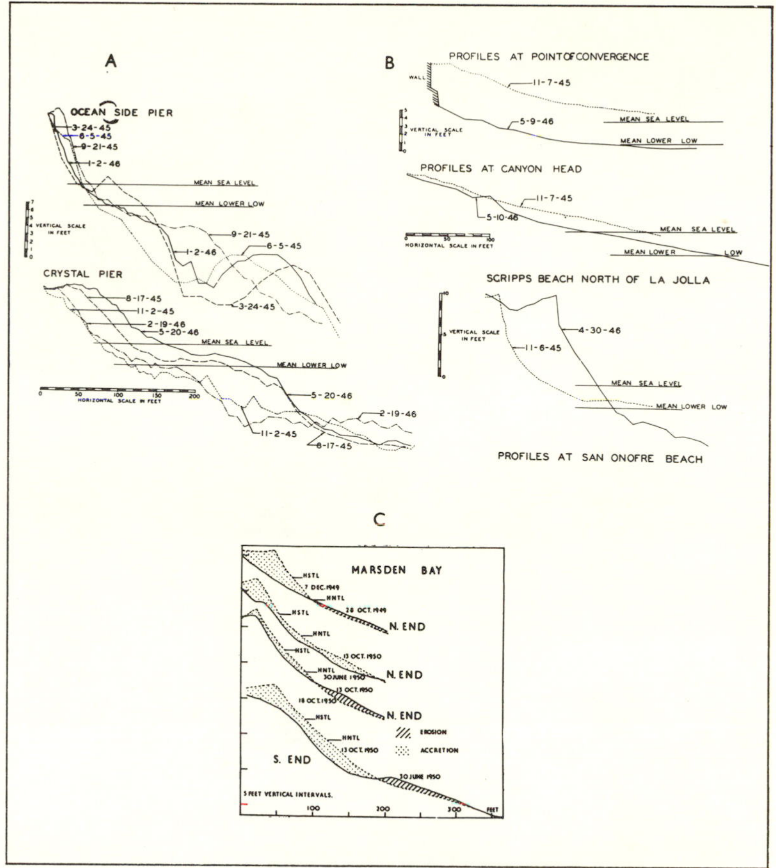

Fig. 6-4. Changes in beaches between winter and summer conditions: (A) along two piers in Southern California; (B) at La Jolla, California, where waves are high (above), and where waves are low (middle)—shows how changes are due to shifting of sand along the beach at San Onofre, California; and (C) along an English beach. (A) and (B) from Shepard, 1948; (C) from C. A. M. King, 1959.

Fig. 6-5. A near vertical scarp off La Jolla. California, cut into the beach berm by large waves.

Fig. 6-6. Boomer Beach at La Jolla, California, showing the (A) summer sand and (B) winter cobbles. During the winter, much of the sand moves south against the low cliffs in the background and returns in summer.

Just around the point to the east of Boomer Beach is the Cove (Fig. 6-7). Here the winter storms with their northwesterly approach have quite a different effect. They push the sand up high onto the shore where there is no lateral escape. In the following summer, the gentle waves at the Cove gradually return the sand to where it was during the previous summer and fall. Thus, here we have an all-year beach.

Types of Beaches

There are three principal types of beach. We have already referred to examples of each. The *small beaches* fitting into a rock indentation, like La Jolla Cove, are located along many headlands. The *long beaches*, which represent the early stage of development of a barrier island, are another type. A third, the *cliffed beach*, is found along the comparatively straight base of a sea cliff and is often traceable for miles (Fig. 6-8). Many years ago, the cliffed beach north of Scripps Institution of Oceanography in La Jolla was proposed as a future site of the railroad. Fortunately, a far-seeing naturalist, Guy Flemming, realized that beaches are ephemeral and may disappear in winter, so with great difficulty he persuaded the city officials to choose an inland route, saving one of the most beautiful beaches in the state of California.

Fig. 6-7. Cove beach in La Jolla, California, where the coarse sand is pushed up onto the beach during winter storms exposing rocks in the low-tide area. Compare with Fig. 6-6 taken in the adjacent area around the corner. The concretions in the sandstone rock in the foreground have persisted with no appreciable change since 1890.

Barrier beaches, a kind of long beach, initially have a lagoon on the inside but, as time goes on, the lagoon changes, first into a marsh and then into a lowland. The filling is partly the result of sediment being introduced by

Fig. 6-8. Part of the extensive beach at the base of the cliffs north of La Jolla, California. Photo taken at low tide, exposing rip current channels formed at high tide.

rains or streams, partly due to plant growth, and partly due to artificial filling to create new real estate.

Beaches can also be classified by texture—*fine sand,* or *coarse sand and gravel.* You don't need to make an analysis of the sand to distinguish coarse from fine. The coarse-sand beaches have a steep foreshore and the fine, a gentle foreshore (see, for example, Figs. 6-1B and 6-7). In fact, there is a fairly good relationship between steepness and grain size, with gravel beaches as steep as 30°, typical coarse sand about 7°, and fine sand about 2°. The fine-sand beaches usually have berms that are nearly horizontal, coarse-sand and gravel beaches generally have berms sloping shoreward at a low degree. Walking is far easier on a fine-sand beach because the sand compacts and usually will hold your weight, in contrast to the coarse sand where you sink in. Fine-sand beaches are often sufficiently compacted to allow cars to run along them, especially on the foreshore. As a result, a few fine-sand beaches are used for auto racing, as at Daytona, Florida.

Some fine-sand beaches become so saturated with water that a person might sink into them in an alarming way. These are called *quicksands* and are supposed to be very dangerous. Having watched young Frenchmen allow themselves to sink into the quicksands at Mont-Saint-Michel, on the Normandy coast of France, I am skeptical of the danger. After they had sunk waist deep, they easily extricated themselves by bending forward to distribute their weight on the sand, and kicking their legs, they emerged with no trouble. After all, one's specific gravity is less than that of a mixture of sand and water, so if you even keep quiet, you can't sink in much beyond your waist. Some panic-stricken people may unwittingly work their way down under quicksand by struggling.

Special Beach Features

While walking along a beach you can see many interesting features that are overlooked by most people. Beach cusps formed by the waves at the shoreline are particularly common in the coarser sand and gravel beaches (Fig. 6-9A). These small points projecting toward the sea are often evenly spaced, their separation depending largely on the size of the waves that produced them and the

Fig. 6-9A. Steep gravel beach with closely spaced cusps.

coarseness of the sand. Well-separated cusps several hundred feet apart are found in fine-sand beaches after large storms, whereas closely spaced cusps a few feet apart are the result of small waves and are more commonly related to coarse sand. Cusps are usually the result of water piling up onto the berm edge and depositing some of the sand as the water sinks into the berm, while the remaining water runs off on either side and returns seaward, causing small indentations. They are formed more commonly where the tides are small or nonexistent, as in lakes or during neap tide periods. Large tidal ranges eliminate them.

Ripple marks are exposed on beaches at low tide. Fine-sand beaches often have ripples of low amplitude on the

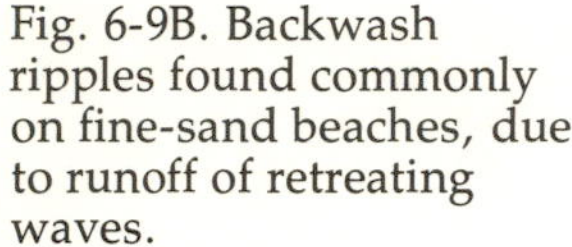

Fig. 6-9B. Backwash ripples found commonly on fine-sand beaches, due to runoff of retreating waves.

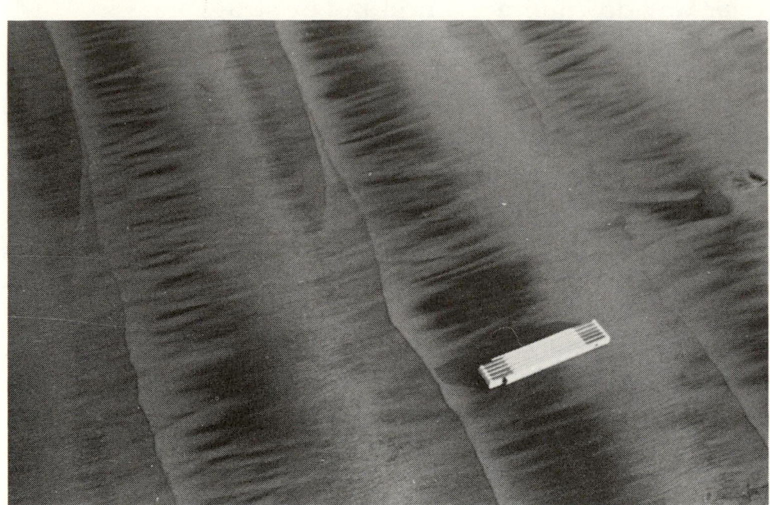

foreshore, separated by about 18 inches (45 cm) and extending parallel to the shoreline (Fig. 6-9B). They are called *backwash ripples* and result from the backwash of waves running down the beach slope. At very low tides the longshore troughs as well as channels due to rip currents may be exposed (Fig. 6-8). Here we see small, closely spaced ripples extending at right angles to the trough channels (Fig. 6-9A). They are steeper on the side toward which the current flows.

Other minor features include *backwash marks* (Fig. 6-9D) due to an obstacle that deflects the downbeach flow, leaving a chevron-shaped feature, often with a concentration of dark-colored sand. Another feature shown at low tide is called a *rill mark* (Fig. 6-9E). These are the result of water

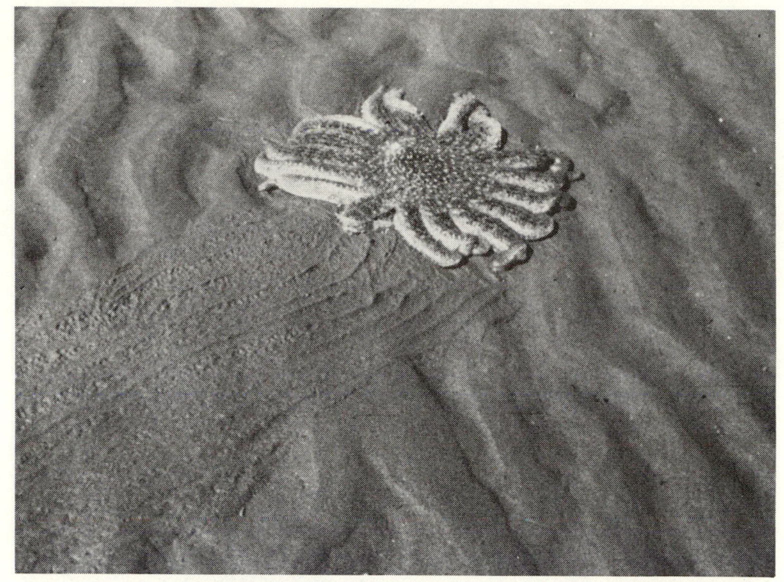

Fig. 6-9C. Current ripples exposed at low tide in channels. Note trail on sand made by the large starfish (*Pycnopodia helianthoides*).

running out of the saturated sand when the tide is low and the water collects in small rivulet channels. Animals, especially birds, leave all sorts of trails and footprints on the sand (Fig. 6-9C). Some of these are preserved when a beach is covered by other formations and later exposed in sedimentary sandstones. Some geologists, notably Hans Reineck (1972) and Adolf Seilacher (1964) of Germany, have made extensive studies of these trails and imprints.

Beach sands are usually stratified, as you can see where

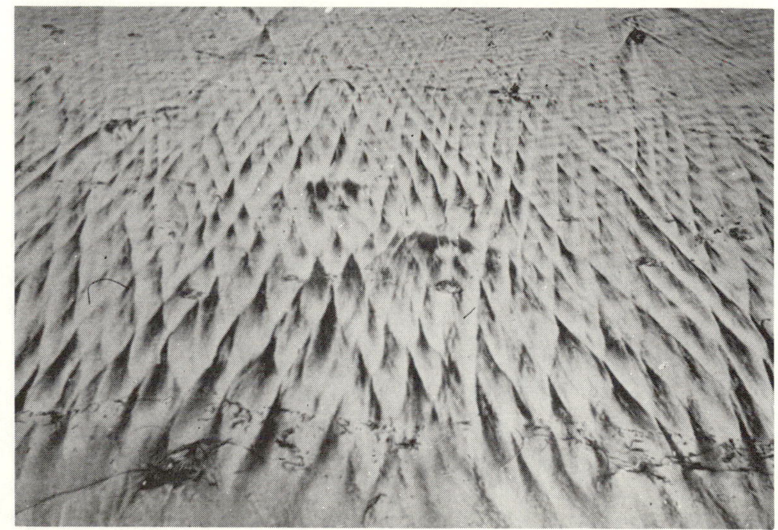

Fig. 6-9D. Backwash marks on a fine-sand beach made by projecting shells, pebbles, or antennae of small sand crabs, which deflect the backwash of waves.

a channel had been cut across them by the rain. Dark layers ordinarily alternate with light, the former having many iron-bearing minerals such as magnetite and ilmenite. These are concentrated during times of powerful wave action. Shell layers in the sand are usually due to small waves that can carry only the small empty shells, such as those of the clam called *Donax*.

Fig. 6-9E. Rill marks due to water running off the beach at low tide, producing small channels.

How Beaches Are Affected by Man

As indicated at the beginning of the chapter, man may be the chief enemy of beaches, just as the human species is its own chief enemy and may eventually destroy civilization. Dams built along river courses are certainly cutting off a large supply of sand to beaches. Flood-control basins along the wet-weather river courses take away still more of the sand, depositing it in the overflow basins. As yet no solution has been developed to stop this loss. We will probably wait until the beaches get seriously depleted before we do anything about them.

Harbor Jetties and Breakwaters. Ever since ancient Mediterranean civilizations developed ships, man has been building jetties out from the shore and breakwaters parallel to the shore to allow ships to anchor and discharge troops and goods in protected areas. Until this century, little thought was given to the effect of these engineering structures on beaches. One of the first serious problems developed when a harbor was built at Santa Barbara, California. Here, the local citizens insisted on establishing a harbor despite warnings from the U.S. Corps of Army Engineers. A few years after the jetty and breakwater had been completed, the hotel owners and residents located east of the harbor began seeing their beaches disappear, and beaches were eventually denuded for 10 miles (16 km) farther east. What had happened was that the sand that had formerly been carried east along the coast was now being trapped behind the harbor jetty (Fig. 6-10A and B), so that to the east the sand cut away during winter storms and carried farther east had nothing to replace it during the following spring and summer. Investigations by the Army Engineers and a few university scientists showed

Fig. 6-10A. The coast at Santa Barbara, California, before the construction of the harbor. Note narrow beach at point west of pier. Photo by U. S. Grant, IV.

many other examples where the beaches downcurrent were being robbed by harbor jetties.

Learning from this example, the harbor at Santa Monica, California, was formed by putting a breakwater parallel to the shore so sand could move through it. Here trouble developed because the breakwater reduced the waves and stopped the sand transport. A wide beach formed inside the breakwater, filling much of the harbor and capturing the sand that was supposed to have fed the downcurrent beaches.

Both at Santa Barbara and Santa Monica the central solution proved very expensive. The Army Engineers installed dredges that sucked the sand away from where it was accumulating and transported it by pipeline to the starving downcurrent beaches.

Owners of Southern California shore property again got into trouble when the Army Engineers permitted the city of Redondo to build a jetty and breakwater for another

Fig. 6-10B. The same area after building the jetty and breakwater. Note wide beach at the zone where sand was trapped. Photo by U. S. Grant, IV.

Fig. 6-11. (A) Beach and (B) boat harbor at Redondo Beach, California, with the breakwater that only partly enclosed it but has since been extended. Note accumulation of sand beyond breakwater and erosion on inside next to jetty (C). The pier is at the head of a submarine canyon (D). Oblique aerial photo, May 1957. Later, the breakwater was extended to check shore erosion.

small-boat harbor (Fig. 6-11). A submarine canyon extends toward shore almost to the Redondo pier (D), and the engineers figured that the sand, which was being carried down the coast, was already being lost by falling into the canyon, so the jetty would not harm the beach south of the pier. They forgot an important principle related to submarine valleys. As we have seen in Chapter 3, waves are diminished at the head of a valley but become concentrated on either side (Fig. 3-2). The city of Redondo had valuable shore property just north of the pier and before building the harbor this area received sand supplies from the north to maintain the beach. Now this was cut off by the jetty and the concentrated waves took over. Despite a new sea wall and the dumping of huge boulders along the shore, the waves cut away an entire city block. Desperate measures were needed, and the breakwater was extended far enough south so that what was left of the waterfront was protected from the wave convergence north of the canyon.

How Can We Preserve Our Beaches for the Future

We have seen how beaches are being lost, largely due to the work of man. Let's see what can be done about it. After-the-fact solutions proved very expensive in the three Southern California harbors just discussed. As a first principle, we should be more careful before we build more harbors. The construction of models in wave-producing tanks would have at least avoided the Redondo debacle. The jetty should have been built farther south so that the breakwater continuation would have protected the shore

Fig. 6-12. The jetties at Newport Harbor, California, which have been quite successful because sources of sand come from both the northwest (upper) and southeast at different seasons of the year. The foreground shows a wave-erosion coast which has become irregular due to differences in resistance of the coastal rock formations. Photo by D. L. Inman.

as far south as the edge of the canyon. Also, it would have become evident from a model study that sand would fill up the Santa Monica Harbor. The Santa Barbara promoters should have established a pumping system to protect the downcurrent beaches, or refrained from making a harbor at that point.

Fig. 6-13. Showing the effect of the groins built at right angles to the beach near Santa Monica, California. The sand drifts to the east (right) and piles up against each groin. Spence Aerial Photo, 1932.

Now let's see if some jetty-protected harbors have been constructed without serious loss to beachfront. Newport-Balboa, California (Fig. 6-12), appears to have been at least reasonably successful. Here we have a location where the large bend in the coast, south and east of the Palos Verdes Hills, has produced currents that flow north during the summer months, providing sand for the beaches south of the jetties; but during the winter, a supply of sand from the Santa Ana River is moved south, feeding the beach at Newport and Balboa. What may be needed in the future is a measure to keep the Santa Ana River as a continuing source of sand by, for example, excluding any additional dams.

Miami and Waikiki Beaches. Perhaps the two best-known American beaches are in Miami, Florida, and Waikiki, Oahu. As beaches, both are serious problems. You get an expensive hotel room in Miami and go down for a swim, and unless the tide is very low, you may find there is no beach. Waikiki is not quite as bad, but the beach in places is so narrow and densely populated that there is often standing room only. What has happened here? In both cases, there were fine beaches in the past, but greedy real estate and hotel developers kept building their houses and hotels farther and farther out on the sand berm so that natural equilibrium conditions were destroyed. As the owners saw their beachfront disappearing, they put in more and more groins (small sea walls vertical to the coast) (Fig. 6-13). Then, when this construction robbed their neighbors downcurrent, they began carrying in sand from other areas. Even this has not saved Miami beaches, where the natural source of sand is from the north and scores of other groins are catching the sand supply upcurrent at Hollywood and Fort Lauderdale. The future for Waikiki is also bleak, because there are not many more beaches in Hawaii that can be stripped for sand supply.

Offshore Sources of Sand. We may have one neglected source of beach sand, and at least for a long time this could save many of our beaches from extinction, because sand is often found in abundance on the continental shelf seaward of beaches. Off the east coast of the United States there are extensive sand areas that are being considered for road building. Why not for beaches instead?

Perhaps if we can find out why the beach to the north of La Jolla, California, has been growing wider, we might get some helpful ideas about other beaches. There is yet a lot to learn.

References

King, C. A. M., 1959. *Beaches and Coasts.* Edward Arnold, London, 403 pp.

Reineck, H. E., 1972. "Tidal flats." In *Recognition of Ancient Sedimentary Environments.* J. K. Rigby and W. K. Hamblin, eds., *Soc. Econ. Paleontol. and Mineralog.*, Spec. Publ. No. 16, pp. 146-159.

Seilacher, A., 1964. "Sedimentological classification and nomenclature of sedimentary structures." *Sedimentology,* v. 3, pp. 253-256.

Shepard, F. P., 1948. *Submarine Geology.* Harper & Row, New York, 348 pp.

Suggested Supplementary Reading

Bascom, W., 1964. *Waves and Beaches: the Dynamics of the Ocean Surface.* Anchor Books, Doubleday and Co., Inc., Garden City, N.Y., Chs. 9-11.

Ingle, J. C., Jr., 1966. *The Movement of Beach Sand: an Analysis using Fluorescent Grains.* Developments in Sedimentology 5. Elsevier Publ. Co., N.Y., 221 pp.

Inman, D. L., and B. M. Brush, 1973. "The coastal challenge." *Science,* v. 181, no. 4094, pp. 20-32.

Johnson, D. W., 1938. *Shore Processes and Shoreline Development.* John Wiley & Sons, Inc., N.Y., 584 pp.

King, C. A. M., 1972. *Beaches and Coasts.* 2nd Ed., St. Martin's Press, N.Y., Chs. 8-14.

Reineck, H. E., and I. B. Singh, 1973. *Depositional Sedimentary Environments with Reference to Terrigenous Clastics.* Springer-Verlag, N.Y., 439 pp.

Shepard, F. P., 1973. *Submarine Geology.* 3rd Ed., Harper & Row, N.Y., Ch. 7.

Steers, J. A., 1969. *Coasts and Beaches.* Oliver & Boyd, Edinburgh, 136 pp.

Changing coastlines in historical times

Chapter 7

We have seen that both marine and land processes change coastlines, either building them out or cutting them back, and that diastrophism (with its attendant earthquakes and the activity of volcanoes) is also effective. How much change has taken place in historical times and how fast are these changes occurring? One hears of ancient cities and even continents lost to the sea. How much of this is fable and how much has a good foundation? The answer is not very clear. As we mentioned previously, the story of the Lost Continent of Atlantis is now generally believed to have originated with the destruction of the Minoan civilization by a great volcanic eruption and a tsunami in the Aegean Sea (Bascom, 1976). But, unfortunately, scientists have made very little attempt to investigate records of the changing coastlines of Eurasia and Africa and verify this belief. In the United States, we have done a little better, but there are very few records from before Colonial times to investigate.

A few years ago an attempt was made to gather together a history of the changes that have taken place along the coasts of the United States (Shepard and Wanless, 1971). Old maps, some going back to the 18th century, were used for the earlier days, and old photographs for locating some changes during the past 100 years were studied. The accurate charting by the U.S. Coast and Geodetic Survey (now called NOAA) of our coasts over the last century was even more helpful. Comparisons were made of aerial photos taken at frequent intervals beginning about 1930 by the Coast and Geodetic Survey and the Department of Agriculture. The results of these studies of the United

States coasts have shown that the major changes have occurred where rivers are building out deltas, where glaciers are retreating, where waves are cutting back coasts of soft material, and where storms are modifying barrier islands and sand spits. Where evidence is available from other countries, the greatest changes come from the same causes, particularly with regard to deltas, of which records are available from relatively ancient times.

Deltaic Modification of Coasts

The largest well-documented coastal change is in the Persian Gulf, where the Tigris-Euphrates Rivers have extended their delta for 188 miles (302 km) since 325 B.C. (Berry et al., 1970). A great advance has also been recorded at the two mouths of the huge Yellow River in China, but the evidence is not as clearly documented. The Nile Delta has been well mapped at least as far back as the Roman Empire, and it seems to have advanced very little. The same is true of the great deltas on the two sides of the Indian Peninsula. The investigations of Morgan and MacIntyre (1959) show clearly that the enormous amount of sediment introduced by the Ganges and Brahmaputra Rivers is being carried offshore, and, as shown by Curray and Moore (1971), is being distributed along the Bay of Bengal rather than building out the deltafront.

For more than a century, American surveyors have been keeping good records of the Mississippi Delta. Figure 7-1 shows how the land was extended out into the Gulf of Mexico between 1838 and 1940. The largest advances occurred where the levees were breached at Main Pass and at the juncture of South and Southwest Passes. Here, the sediment-laden water flowing through the gaps built rapidly into shallow water, forming vast marshlands between the many distributaries. About 10 miles (16 km) of new land was added at Main Pass, and a little less to Johnson Pass, where Garden Island Bay has been largely filled, although here the shorelines between 1949 and 1956 show almost as much retreat as they do advance (Fig. 7-2). The investigation of Morgan (1963) covering the entire Mississippi Delta (which includes all of the passes back to

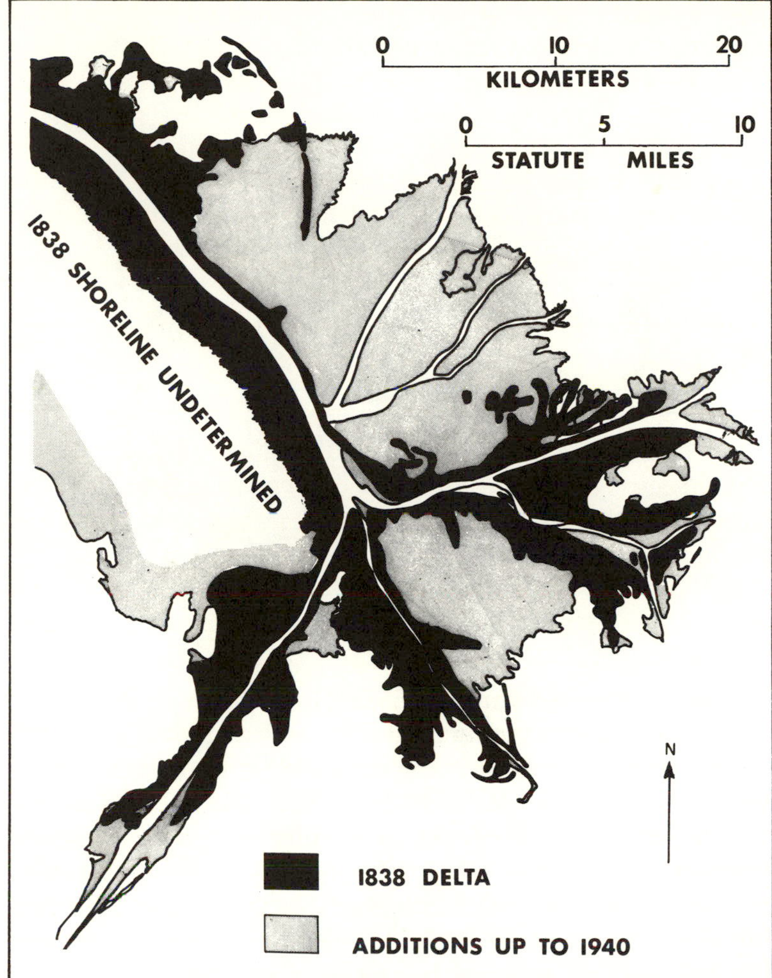

Fig. 7-1. Showing the growth of the Birdfoot Delta of the Mississippi during a 100-year period. From U.S. Coast and Geodetic Survey charts.

about 5,000 years before the present) shows that in recent years there has actually been slightly more loss of land than gain, so instead of Louisiana growing larger it actually has less land area than indicated in the earliest surveys. This is due to the general sinking that effects most large deltas and to the slowing of the deltafront advance where it has built seaward into relatively deep water.

Along the rest of the coast of the United States there is clear evidence of presently advancing coastlines only where rivers are entering embayments. Next in size to the Mississippi among American deltas is the Yukon. This has a lobe that has been built into the Bering Sea (Fig. 7-3). However, a chart made in 1898 shows no evidence that this lobe has been growing larger in recent years. It pre-

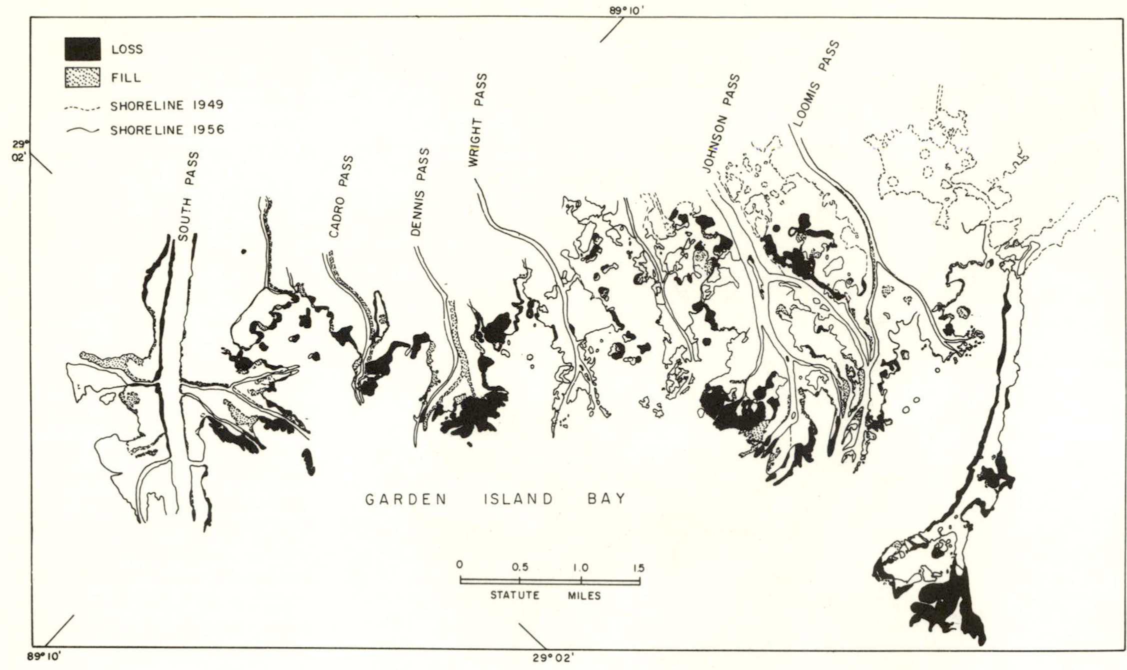

Fig. 7-2. The balancing of loss and fill of the passes entering Garden Island Bay, east of South Pass of the Mississippi Delta, between 1949 and 1956. From U.S. Coast and Geodetic Survey charts.

sumably grew forward after sea level stabilized but has now stopped advancing. The Brazos River, the largest in Texas, may have an advancing delta. It had built out almost a mile (1.6 km) between 1858 and 1958, but the mouth was then diverted artificially and the advanced portion of the delta was partly cut away by hurricanes. Similarly, the present mouth is advancing largely by building of arcuate barriers (Fig. 7-4), leaving a lagoon that was later filled. However, hurricanes interfere with this growth and we cannot be sure that the coast is being permanently increased.

Sediment from rivers can build rapidly into an embayment. Thus, the Colorado River of Texas was discharging into Matagorda Bay in 1831 when a raft of logs blocked the mouth and caused deposition upstream. After the logjam broke during a flood in 1929, the river began a rapid advance, building a delta five miles (8 km) across the bay to the barrier island by 1950. The U.S. Army Engineers cut a ditch through the island to allow the water to flow out into the open Gulf (Fig. 7-5). There is no sign of a new delta advancing into the Gulf since that time. A less rapid growth into a bay is seen where the Trinity River is building into Galveston Bay and has advanced about a mile (1.6 km) since 1855.

Retreating Glaciers Forming New Embayments

The fastest average retreat of coastlines in historical times seemed to have occurred in Alaska. According to the charting in 1794 by Captain Vancouver, there was a tongue of ice protruding out into the open sea from Malaspina and Guyot Glaciers. This has now retreated almost 40 miles (64

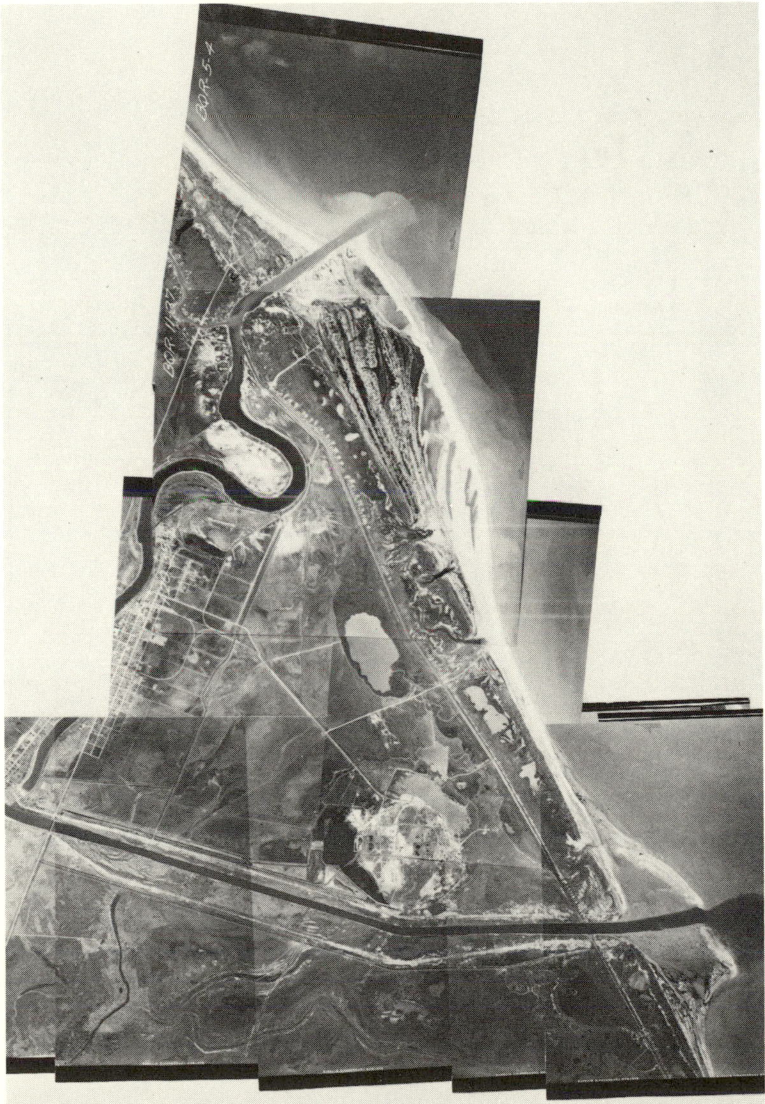

Fig. 7-3. The new and old outlets of the Brazos River of Texas. After artificial deflection of the river into the new channel, the delta of the old channel (right) had been largely cut away and the new delta had been started by a cuspate bar outside the new channel (left). Aerial photo from U.S. Department of Agriculture, 1938.

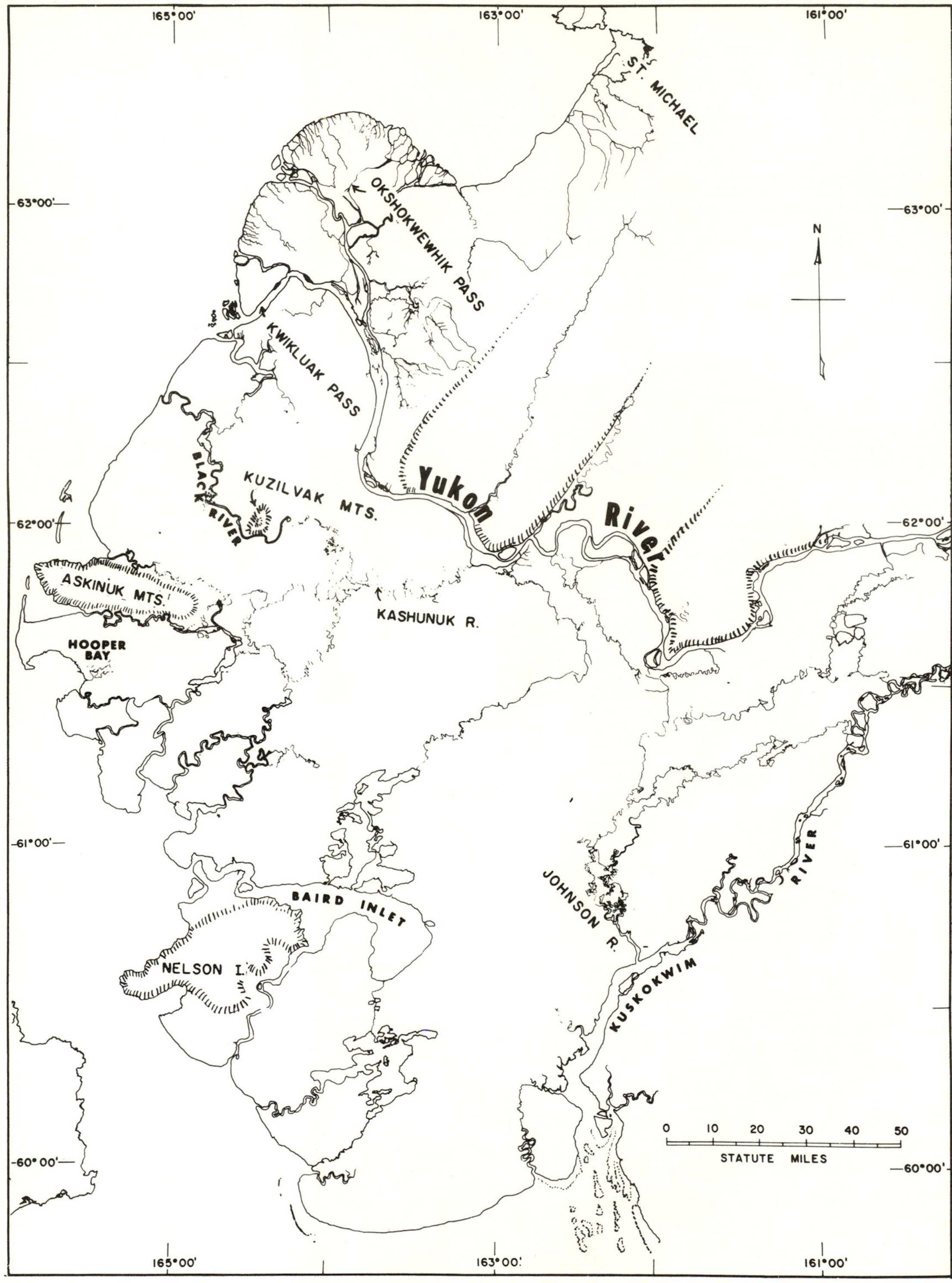

Fig. 7-4. The Yukon Delta built into the Bering Sea, Alaska. The Okshokwewhik Pass to the north, which carries most of the water to the sea, has not grown appreciably since 1898. Older outlets are shown to the south. From Shepard and Wanless, *Our Changing Coastlines* (New York: McGraw-Hill, 1971).

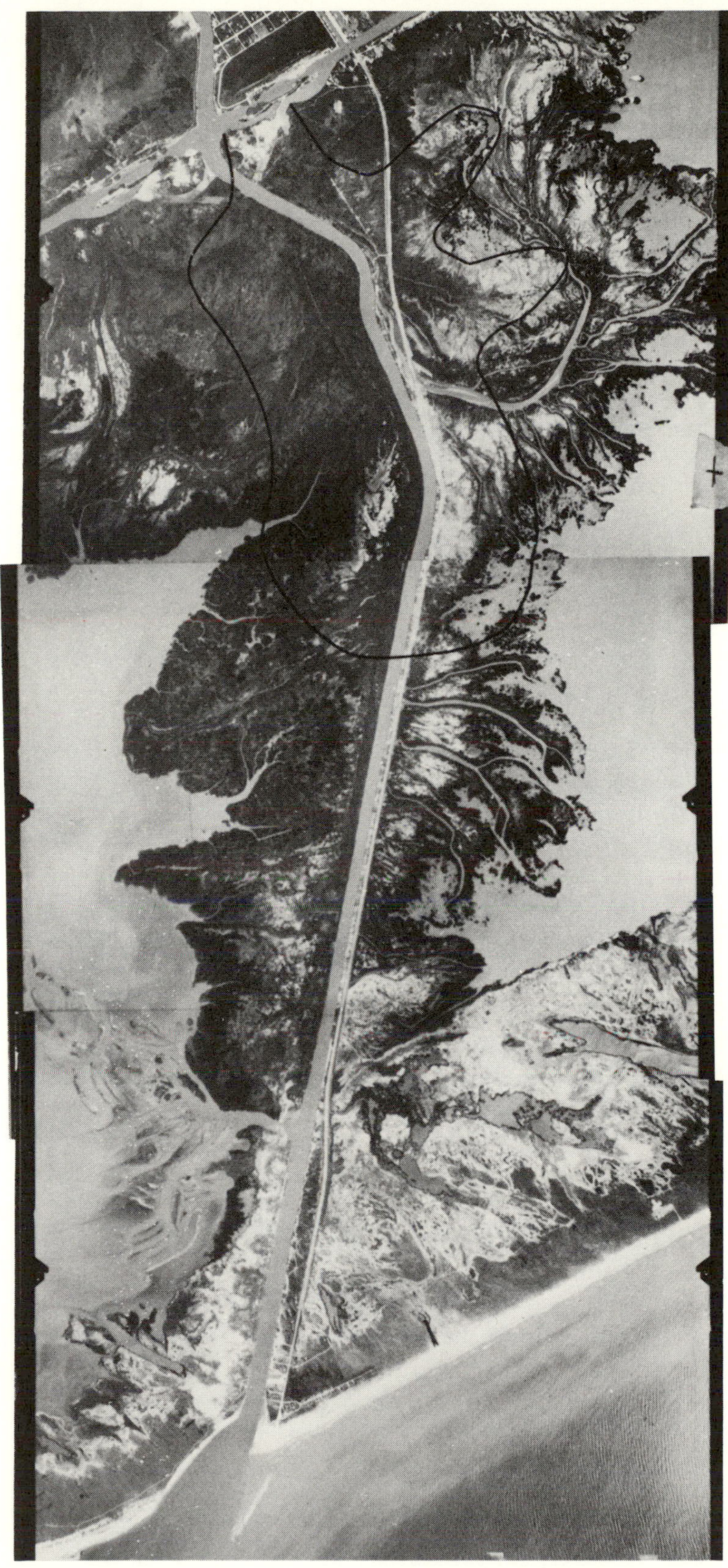

Fig. 7-5. The delta of the Colorado River of Texas. The black line shows the delta edge in 1930, a year after the rapid advance had begun due to the breaking up of a log jam in 1929. By 1950, the delta had built 5 miles (8 km) across Matagorda Bay, and the U.S. Army Engineers had to cut a pass through Matagorda Island to allow the river to enter the Gulf of Mexico. From U.S. Department of Agriculture, 1955.

km) in less than 200 years, exposing Icy Bay (Fig. 7-6). Perhaps one of the largest retreats is found in Glacier Bay, located 50 miles (80 km) west of Skagway, where gold miners used to take off for the Klondike gold fields. Glacier Bay was almost full of ice in 1794, according to Vancouver. It has now retreated some 60 miles (97 km). According to the maps of the Coast and Geodetic Survey, a retreat of 7 miles (11 km) has occurred in the past 30 years. The present Johns Hopkins Glacier is calving many large icebergs into the fiord and may still be retreating.

It is possible that even larger changes of coastlines have occurred during the past two centuries in Antarctica. Since aerial mapping began about 1930, some notable retreats of the ice have been shown, especially at Little America.

Cutting Back of Cliffs by Waves

How fast do sea cliffs retreat? That is an important question if you are interested in building near the cliff edge where you get the best ocean view. A precise answer is far from easy to get. For example, extending south from Scripps Insitution of Oceanography at La Jolla, there was a line of low vertical sea cliffs cut into soft alluvium (Fig. 7-7A). These were cutting back at a rate of approximately one foot (0.3 m) per year. Hence, one would have warned builders to follow the example of the Scripps Institution and put in sea walls before building near this cliff edge. However, the retreat of the cliffs virtually stopped in 1947 when, as discussed in the previous chapter, the berm ceased to be cut away by winter storms. Now there are many houses built along the slumped remnants of the old sea cliff with no protection (Fig. 7-7B). Even one of the new Scripps buildings is unprotected at the north end of the old alluvial cliffs. Unfortunately, a renewal of cliff retreat may set in.

Meanwhile, some 10 miles (16 km) to the north at Del Mar and extending up to Encinitas, the 100 to 200-foot (30 to 60 m) cliffs, also cut mostly in alluvium, have been retreating so fast that the new condominiums built near the cliff edge are seriously threatened (Fig. 7-8). Studies by Gerry Kuhn (in preparation) show that retreat has

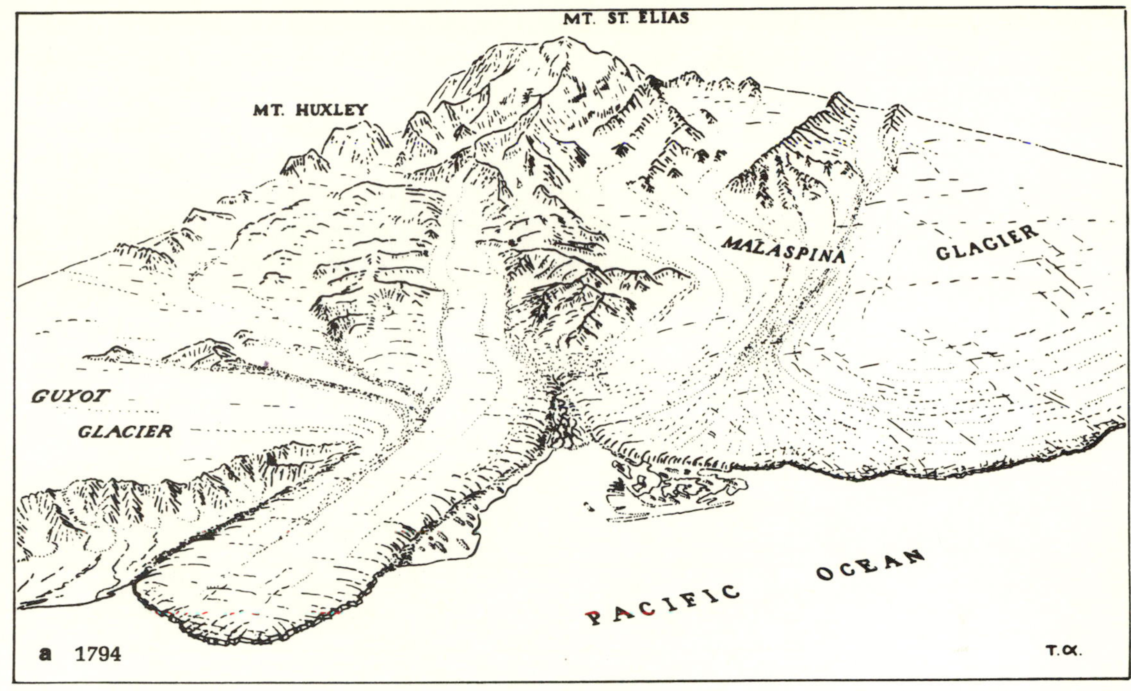

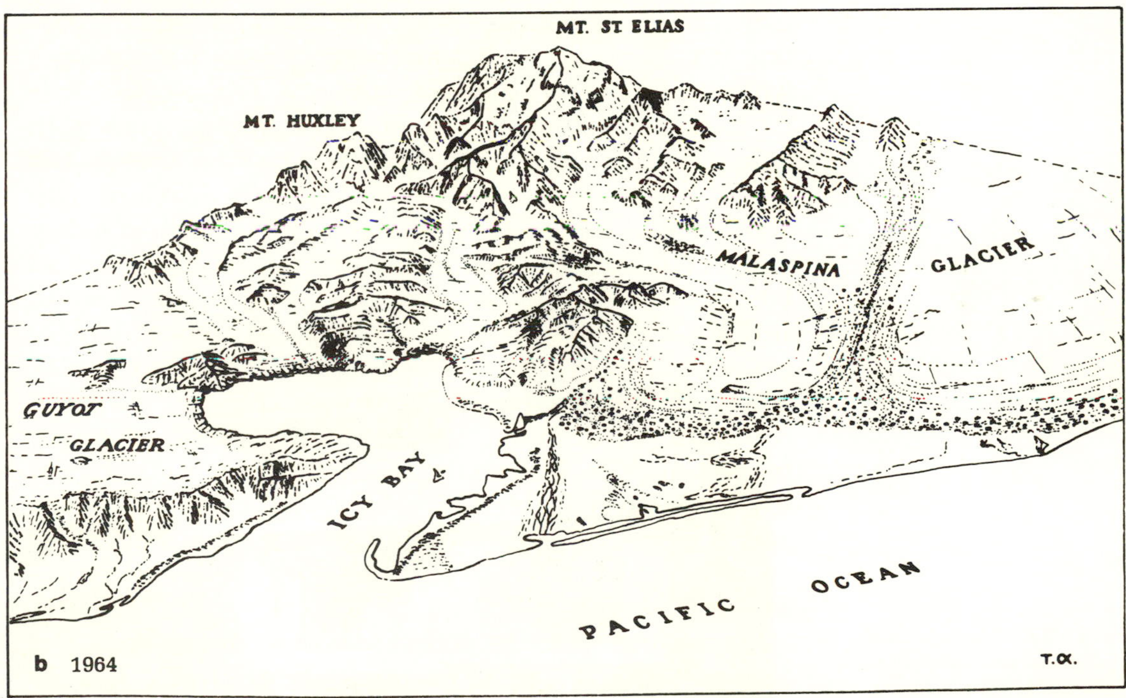

Fig. 7-6. Block diagram showing the 40-mile (62-km) retreat of the Malaspina-Guyot Glaciers since they were sketched by Vancouver in 1794. Icy Bay was uncovered and a large sand barrier was built on its east side. Drawn by T. R. Alpha of the U.S. Geological Survey.

7-7A. After retreat had stopped, at least temporarily, the cliffs slumped to a gentler angle. See also Fig. 6-1.

Fig. 7-7B. Houses built with no sea walls along the cliffs shown in Fig. 7-7A, and, after retreat had stopped, at least temporarily, the cliffs slumped to a gentler angle. See also Fig. 6-1.

Fig. 7-8. Condominiums built at the edge of a rapidly retreating alluvial cliff in Solana Beach, California. Collapse of a sea cave and subsequent bluff collapse formed a fan of debris shown left of center. New scarp is below left building. Photo by Gerry Kuhn.

amounted to as much as 3 feet (0.9 m) per year. A temple of the Self Realization Fellowship built in 1939 fell over the cliff at Encinitas in 1941 (Fig. 7-9A and B).

One of the best recorded examples of wave erosion in unconsolidated material comes from the 150-foot-high (46 m) alluvial cliffs south of Cape Cod Light on the southern Massachusetts coast. During the 19th century, the lighthouse had to be replaced at least three time. Studies by John Zeigler and others (1964) of Woods Hole showed that over a 70-year period the cliff edge had retreated 175 feet (53 m) or 2.5 feet (0.76 m) per year.

Rates of erosion, cited by the English coastal expert C. A. M. King (1972), give many other examples of cliff retreat. She states that in Huntcliff, Yorkshire (an old Roman signal station), a retreat of 100 feet (30 m) has occurred in the past 800 years. Quoting H. Valentin, she refers to the average erosion of 7 feet (2 m) per year along

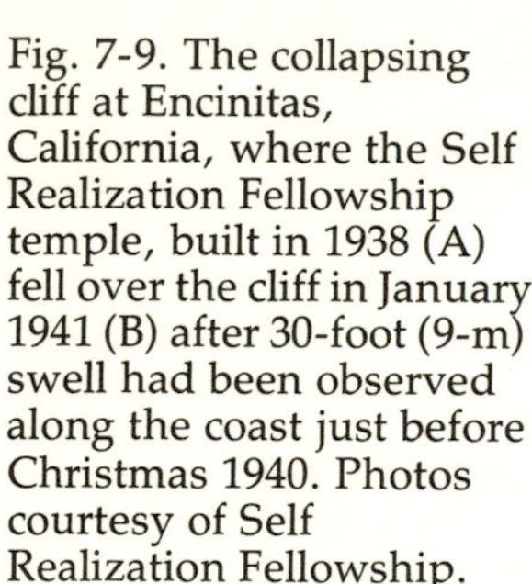
Fig. 7-9. The collapsing cliff at Encinitas, California, where the Self Realization Fellowship temple, built in 1938 (A) fell over the cliff in January 1941 (B) after 30-foot (9-m) swell had been observed along the coast just before Christmas 1940. Photos courtesy of Self Realization Fellowship.

the Holderness Coast; and, quoting J. A. Steers, of 2.6 feet (0.8 m) per year for the soft cliffs of Suffolk. Steers found a rate of erosion of about 13 feet (4 m) per year in East Anglia. The chalk cliffs of northern France, according to Steers, are retreating 0.9 feet (0.27 m) per year. Thus, erosion in soft material can be very fast.

Up to now, we have been discussing cliffs cut in soft, easily eroded material. Soft shale or chalk may erode almost as fast as alluvium, but hard granite erodes so slowly that, for example, the glacial striations uncovered along the Maine coast by the retreat of the ice some 10,000 years ago are seen extending down a granite slope into the sea with no trace of erosion (Johnson 1925).

We have a fairly good collection of photographs taken along the California coast dating back almost a century, which gives some idea of the extent of erosion of firm rock cliffs. Perhaps the most surprising discovery that has come from retaking photographs from the same locations is that the cliffs and even the rock stacks outside the cliffs often look as though nothing had happened (Fig. 7-10A and B). However, a more careful inspection of a locality such as La Jolla, California, shows that there have been occasional landslides that caused the cliffs to retreat a score or more feet. Some of the old landmarks, like the sea

Fig. 7-10. Coastal scene 5 miles (8 km) north of Port San Luis, California. Despite exposure to open sea, the 1945 photo (B) taken by U. S. Grant, IV shows no change from the 1898 photo (A) taken by G. W. Stose.

Fig. 7-11. Two stages in the destruction of Cathedral Arch at La Jolla, California. (A) was taken in 1899 just before the collapse, and (B) in 1938. The 1938 remnant has now completely disappeared.

arch, called Cathedral Rock, at La Jolla (Fig. 7-11A and B) have not only collapsed but the rocks have been completely ground up into sand. On the other hand, the sea arch at the east end of Anacapa Island (as shown in an engraving by the great artist James Whistler in 1854 when he was with the U.S. Coast and Geodetic Survey) looks almost untouched at the present day (Fig. 7-12A and B).

Fig. 7-12. An engraving by the famous artist James Whistler in 1854 of the sea arch at the east end of Anacapa Island off Santa Barbara, California (A). Note that the photograph taken by Patrick McLoughlin in 1976 (B) shows no indication that the arch has changed. The difference in width of passages is due to photo being taken from a point a little to the west of where sketch was made for the engraving.

Changes in Barriers and Sand Spits

The most conspicuous short-period changes of coastlines are found in barrier islands and sand spits. Retaking aerial photos after a hurricane shows that a barrier island can change to an amazing degree (Fig. 7-13A and B). Prior to a large storm, the beach may be continuous for many miles along the sea front of the island; but after the storm a series of channels usually cut the beach, and many inlets extend through to the lagoon on the inside. Sediment from the beach often forms storm deltas in the lagoon. Some of these have extensive flats, adding appreciably to the width of the barrier and partly filling the lagoon (Plate 2, Fig. 1).

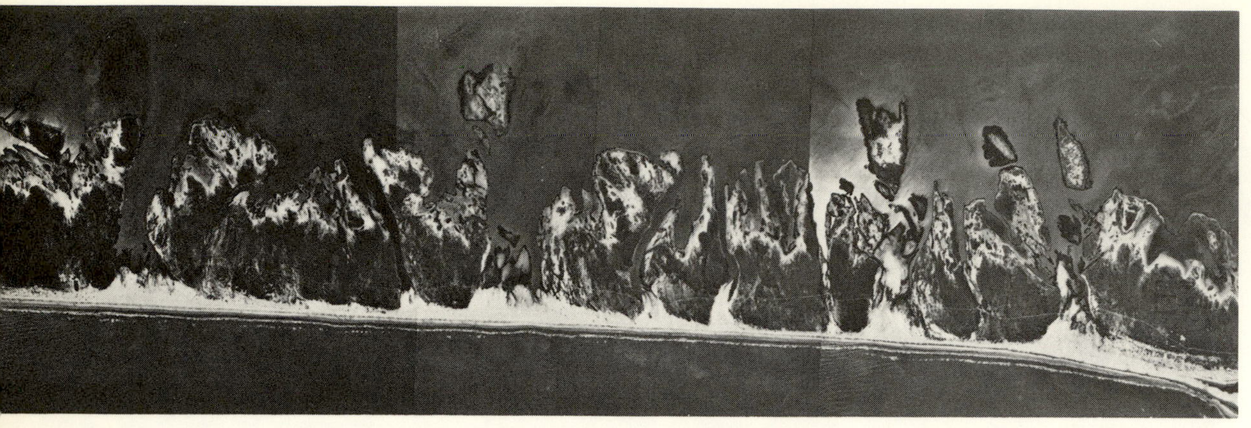

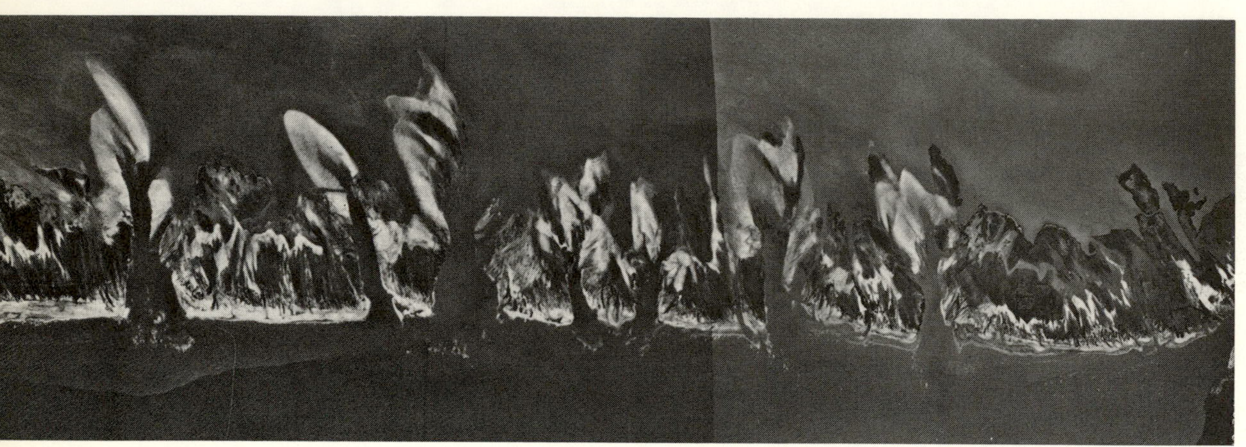

Fig. 7-13. The beach extending along the Matagorda Peninsula barrier in September 1960 (A), and the same area September 17, 1961 (B) after Hurricane Carla had produced a series of cuts through the barrier and removed the beach. From U.S. Coast and Geodetic Survey.

We have an excellent series of maps of the coast forming the low south elbow of Cape Cod that show how the sand that was cut away from the outer cliffs at the north end of Cape Cod has gradually extended the shore southward (Fig. 7-14). The sand from the Cape Cod cliffs may also move northward, the direction depending on the wind and the tide. Beyond the cliffs, the northward flow curves to the west and has built out a sand spit, which in turn forms a hook enclosing Provincetown Harbor (where the Pilgrims made their first landfall in 1620). Unfortunately, they were too much occupied with survival even to sketch a map for posterity showing the nature of the hooked spit for investigators three and a half centuries later. But we do know that the spit has grown slightly in recent years.

Fig. 7-15 shows a good recent history of the changes of the cuspate forelands at Capes Hatteras, Lookout, and Fear. In general, they show a growth of new points on the south side of the capes because of the prevalence of northeast winds during storms. However, when storms have come from the south, the tips of the capes have turned temporarily to the north.

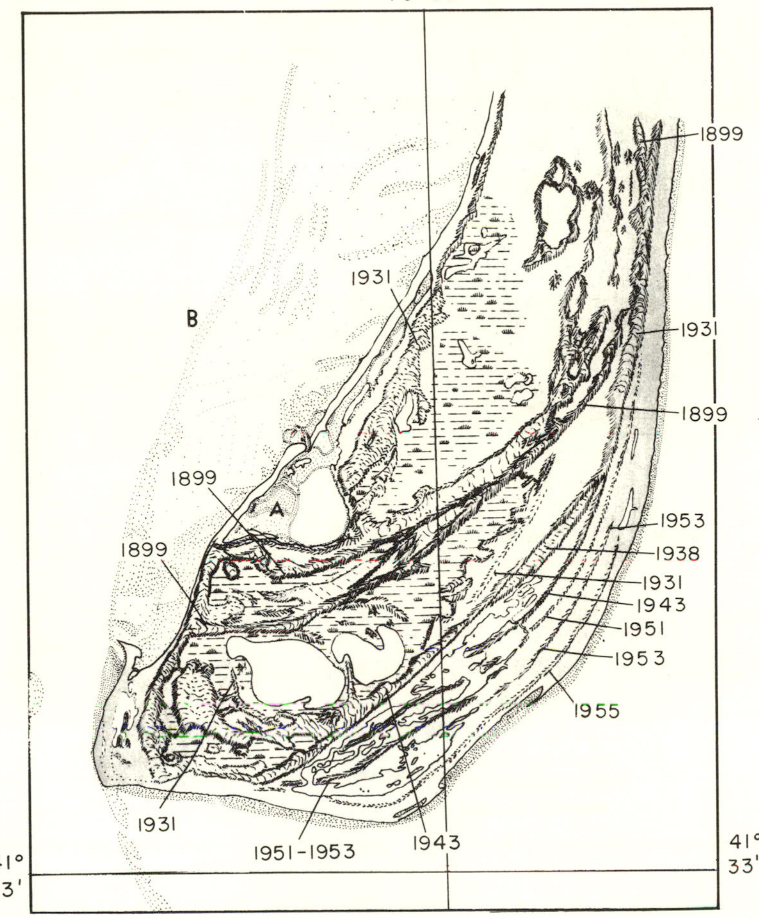

Fig. 7-14. Showing the nature of coastal growth of Monomoy Island at the southeast end of Cape Cod. From various surveys reproduced by Harold Roger Wanless.

Changes Due to Volcanism

The effects of volcanic eruptions on coastlines are not as well documentated as those of the processes considered previously, but we do have some striking examples. Bogoslof, the tiny island in the southeast corner of the Bering Sea, has had an interesting documentation (Fig. 7-16) going back to the Russian explorations of 1769, when

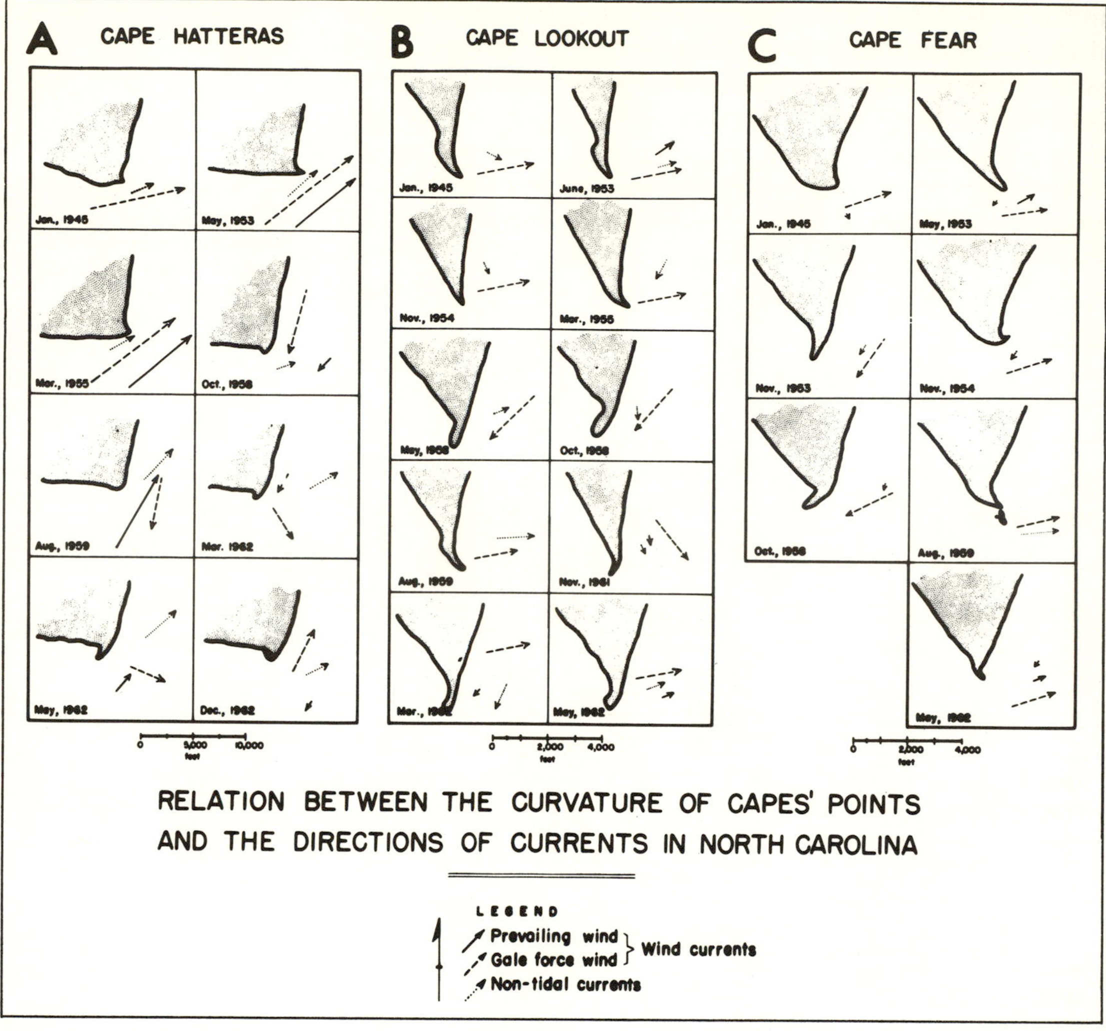

Fig. 7-15. Showing the relationships of the curvature of points at Cape Hatteras, Lookout, and Fear, and the direction of storm winds that preceded the aerial photographs. From U.S. Weather Bureau data compiled by M. T. El-Ashry.

it consisted only of a pinnacle rock. Bogoslof appeared first as a volcano in 1796. Its Russian name, given at that time, *Ioann Bogoslof*, means John the Baptist. The series of sketches from various sources reproduced in Figure 7-16 present the principal changes that occurred to the island up to 1907, as compiled by Thomas Jaggar (1908) of Hawaii. Byers (1959), of the U.S. Geological Survey, brought the history of the island up to 1958 and showed further important changes.

A much more violent modification took place in a larger island, Krakatoa, in 1833, when the island exploded and collapsed, causing the worst tsunami on record. The exact nature of the island before the eruption is unknown, so far as I can determine. Iceland has had important coastal changes from its active volcanoes. In 1973, a new volcano formed, building an island along the south coast. San Benedicto Island off the Mexican west coast was almost completely transformed by the building of a new volcano in 1952 and 1953 (Fig. 7-17). The stages of growth were monitored by Richards and Dietz (1956) with the help of the U.S. Air Force. Many flows have carried lava to the sea in the past century along the south and southwest coasts

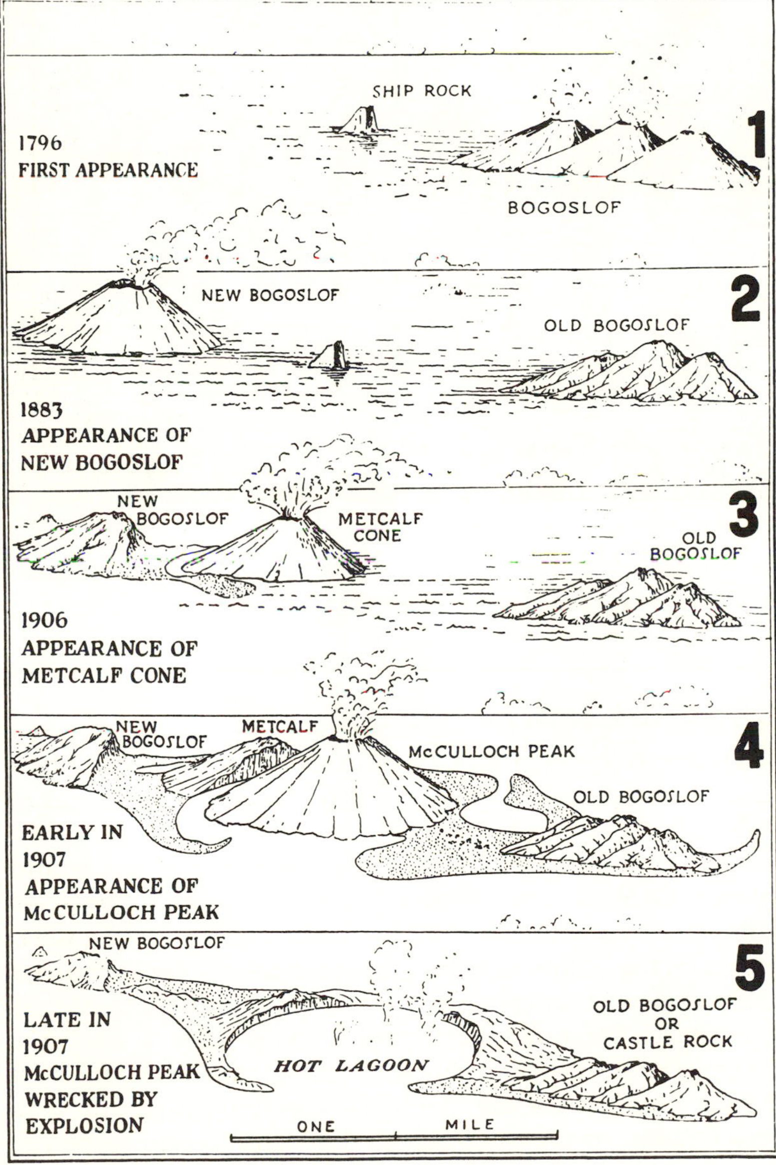

Fig. 7-16. Changes in Bogoslof Volcano between 1796 and 1907. From a compilation by Jagger (1908).

Fig. 7-16. The new volcano Barcena which grew in the middle of San Benedicta Island off western Mexico in 1952 and 1953 as a result of a series of lava flows, ash flows, and *Nuées ardentes* (heavy gas clouds). From Richards and Dietz, 1956.

of Hawaii. These have built out the land to a slight degree, making a few small projections.

Coastal Changes at the Time of Earthquakes

As we saw in Fig. 5-6, a wave-cut bench was brought above sea level at the time of the 1964 Alaska earthquake. The same earthquake was accompanied by land subsidence in many areas, notably at the head of the Turnagain Arm, where 5 miles (8 km) of the railroad right-of-way was inundated near the town of Portage.

Perhaps the most spectacular change was at the Rann of Kutch, south of the Indus River in western India. This area of 9,000 square miles (23,310 sq. km) was an arm of the sea in the 4th Century B.C., but was gradually silted up so that it is now only covered by a thin sheet of water during strong southeast monsoons and is a desert the rest of the time (Krishman, 1960). In 1819, a great earthquake raised the center of this area by several feet and a vast area to the south was submerged. More than a thousand people were killed by the earthquake.

References

Bascom, W., 1976. "Science and ancient sea stories." *Oceans*, v. 9, no. 4, pp. 10-11.

Berry, R. W., G. P. Brothy, and A. Haqash, 1970. "Mineralogy of the suspended sediment in the Tigris, Euphrates, and Shatt-al-Arab Rivers of Iraq, and recent history of the Mesopotamian Plain." *Jour. Sed. Petrology*, v. 40, no. 1, pp. 131-139.

Byers, F. M., 1959. "Geology of Umnak and Bogosloff Islands, Aleutian Islands, Alaska." *U.S. Geol. Surv. Bull.* 1028-L, P. 8, pp. 267-369.

Curray, J. R., and D. G. Moore, 1971. "Growth of the Bengal deep-sea fan and denudation in the Himalayas." *Geol. Soc. Amer. Bull.*, v. 82, no. 3, pp. 563-572.

El-Ashry, M. R., and H. R. Wanless, 1968. "Photo interpretation of shoreline changes between Capes Hatteras and Fear (North Carolina)." *Mar. Geol.*, v. 6, no. 5, pp. 347-379.

Jaggar, R. A., Jr., 1908. "The evolution of Bogoslof Volcano." *Amer. Geogr. Soc. Bull.*, 40, pp. 385-400.

Johnson, D. W., 1925. *New England—Acadian Shoreline.* Wiley, N.Y., 608 pp.

King, C. A. M., 1972. *Beaches and Coasts.* 2nd Ed., St. Martin's Press, N.Y., 403 pp.

Krishnan, M. S., 1960. *Geology of India and Burma.* Higginbothams (Private) Ltd., Madras, p. 576.

Morgan, J. P., 1963. "Louisiana's changing shoreline." *Coastal Studies Inst.*, Louisiana State Univ., Contrib. 63-5, pp. 66-78.

Morgan, J. P., and W. G. McIntire, 1959. "Quaternary geology of the Bengal Basin, East Pakistan and India." *Geol. Soc. Amer. Bull.*, v. 70, no. 3, pp. 319-342.

Richards, A. F., and R. S. Dietz, 1956. "Eruption of Barcena volcano, San Benedicto Is., Mexico." *8th Pacific Congress, Philippines,* v. 2, pp. 157-176.

Shepard, F. P., and H. R. Wanless, 1971. *Our Changing Coastlines.* McGraw-Hill Co., N.Y., 579 pp.

Ziegler, J. M., S. D. Tuttle, G. S. Giese, and H. J. Tasha, 1964. "Residence time of sand composing the beaches and bars of outer Cape Cod." *Proc. 9th Conf. Coastal Engineering,* Amer. Soc. Civil Engrs., pp. 403-416.

Suggested Supplementary Reading

Shepard, F. P., 1976. "Coastal classification and changing coastlines." In *Geoscience and Man,* v. 14 of *Coastal Research,* H. J. Walker, ed., Louisiana State Univ., Baton Rouge, pp. 53-64.

Zenkovich, V. P., 1967. *Processes of Coastal Development.* J. A. Steers and A. M. King, eds. (trans. by D. G. Fry). Interscience Publ., John Wiley & Sons, Inc., N.Y. Chs. 10 and 11.

SANTA BARBARA
SANTA BARBARA BASIN
SANTA MONICA BASIN
SANTA CRUZ IS.
SANTA CRUZ BASIN
SANTA ROSA IS.
SAN MIGUEL IS.
SANTA ROSA
CORTES
TANNER BASIN
PATTON ESCARPMENT
SAN JUAN SEAMOUNT
33°
121°

Continental shelves and their treasure chests

Chapter 8

Much is being written in the press about the vicissitudes of the continental shelves; either that they are the great hope of the Western World in the energy crisis or are a major disappointment. We will probably know before long which is correct, but undoubtedly the continental shelves are underlain by large supplies of petroleum and natural gas. At the present time large quantities of oil are being extracted from the shelf along the Gulf Coast of the United States, in the North Sea, in the Yellow Sea, off the California coast near Santa Barbara, off northern Venezuela, off west equatorial Africa, in the Persian Gulf, and in the Soviet zone of the Black Sea. Sea-floor extraction is developing off both north and south Alaska, on the Bombay shelf, around Indonesia, around Australia in Bass Strait and off its northwest coast, and in the straits on the west side of Taiwan. Innumerable other shelf areas are being investigated, notably the shelf of Georges Bank off New England and off Chesapeake Bay farther south. It may well be that the major extraction of oil in the next century will be from the sea floor, perhaps even from areas out beyond the continental shelves.

How do we define the continental shelf? Geologists and statesmen are not entirely in agreement as to the meaning of the term. There used to be an understanding by Americans and British that the shelf extended seaward from the coast to the 100-fathom (600-ft) depth. Those countries using the metric system referred to the edge as being at either 100 or 200 meters. This type of definition was most inadequate, a fact which became evident when President Truman annexed the mineral rights of the U.S. continental shelves in 1945. Examination of charts of the world shows that off most coasts the sea floor slopes gently seaward until there is a rather sudden steepening in slope that leads down to the deep ocean floor (Fig. 8-1). The average depth of the change is about 67 fathoms (123 m). However, the change is found locally at depths as great as 500 fathoms (914 m), and elsewhere at less than 10 fathoms (18 m). The use of 100 fathoms (183 m) as the shelf edge is most unsatisfactory where glaciers have carved out great valleys deeper than 100 fathoms (183 m) right next to the coast, as in southern Norway and in the Gulf of St. Lawrence.

International conferences of marine geologists have attempted to define the continental shelf as the principal break in the gentle seaward slope that occurs at a depth of less than 300 fathoms or 600 meters, according to which system is used. The inclined floor beyond the shelf edge we call the *continental slope*, except in its lower portion where there is a distinct decrease in slope that is called the *continental rise*. Difficulties arise from the large number of small new nations with no sea coast. They want to share the wealth to be anticipated from the development of what was formerly considered international territory when we had 3- and 12-mile national limits of sovereignty. It seems likely that the resources of the continental shelves will be given to the maritime nations, but the deep sea (to be discussed in Chapter 11) may become international property. The great difficulty seems to be that the maritime nations are the only ones that have the capital to develop these marine resources (Wooster, 1976).

Types of Continental Shelf

There are several very distinct types of continental shelf. These are mostly related to the climate of the adjacent land and presumably to the effects of plate tectonics. The result of the changing sea levels associated with growth and melting of the great continental glaciers has, of course, been of importance to all continental shelves.

Glaciated Shelves. The ice caps that spread from three or

more centers in Canada, one center in northern Europe, and a center in southern Argentina, all encroached on the continental shelves to some extent (Figs. 8-2, 8-14). This statement is based on the character of the shelves adjacent to most glaciated land masses. At first, the troughs in the shelves with water several hundred fathoms deep were thought to be the result of faulting (Gregory, 1920). The

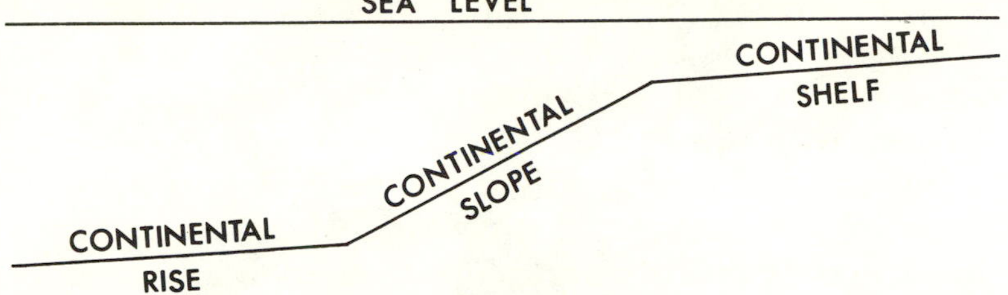

Fig. 8-1. Profile showing the relationships of continental shelf, slope, and rise.

deep trough with its steep sides that comes out of the Gulf of St. Lawrence (Fig. 8-3) was attributed by several geologists to the faults that caused the 1929 Grand Banks earthquake and its cable breaks (Keith, 1930). Douglas Johnson (1925) described a fault extending along the Maine coast as coming from the Bay of Fundy; and Olaf Holtedahl (1940), of Norway, cited many faults extending across and along the shelves bordering Scandinavia. But these interpretations failed to account for the fact that almost all glaciated areas showed the same straight, elongated, steep-sided depressions, and that many of these troughs are clearly an extension of glacial fiords. This is no coincidence. Just as glaciers scooped out troughs on land (leaving lakes, like the Great Lakes and the lakes of Switzerland), they accomplished the same thing where they crossed the shallow sea floor. (We now know that in some places the glaciers did cut along old faults so that the earlier interpretations were not entirely wrong.)

Another effect of glaciation on the continental shelves was the formation of shoals, in many cases on the outer shelves. Thus we have the Grand Banks off Newfoundland (Fig. 8-4) and Georges Bank off the New England coast (Fig. 8-5). Off Nova Scotia, one of these banks extends above the surface, forming wind-swept Sable Island. Georges Bank comes within a few feet of the surface at Georges Shoal. However, most of the banks off the fiord coasts of Norway and British Columbia are considerably deeper. All these banks have huge fish supplies, al-

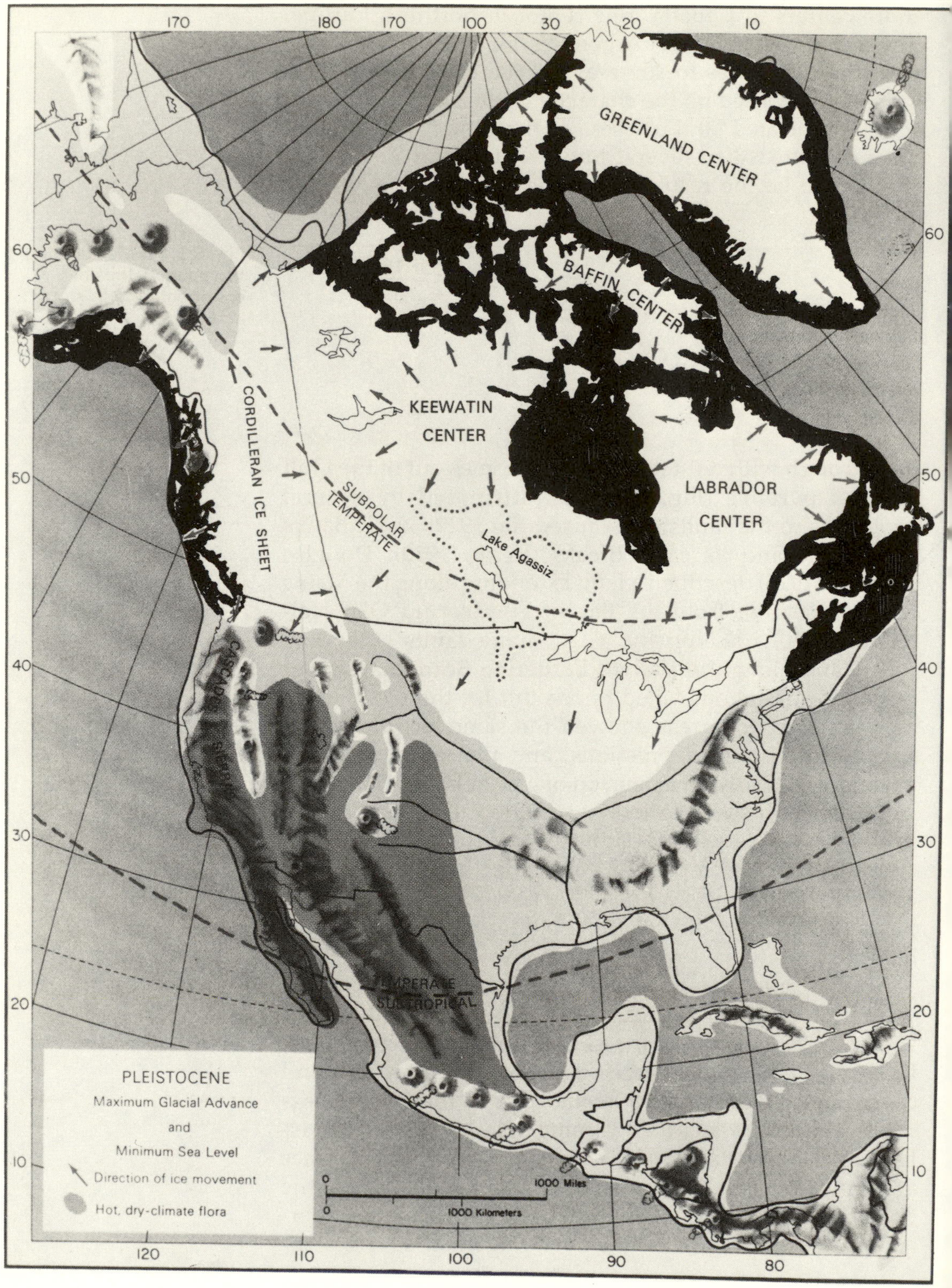
GREENLAND CENTER
BAFFIN CENTER
KEEWATIN
CENTER
LABRADOR
CENTER
CORDILLERAN ICE SHEET
SUBPOLAR
TEMPERATE
Lake Agassiz
CASCADE
SIERRA
TEMPERATE
SUBTROPICAL
PLEISTOCENE
Maximum Glacial Advance
and
Minimum Sea Level
Direction of ice movement
Hot, dry-climate flora
1000 Miles
1000 Kilometers

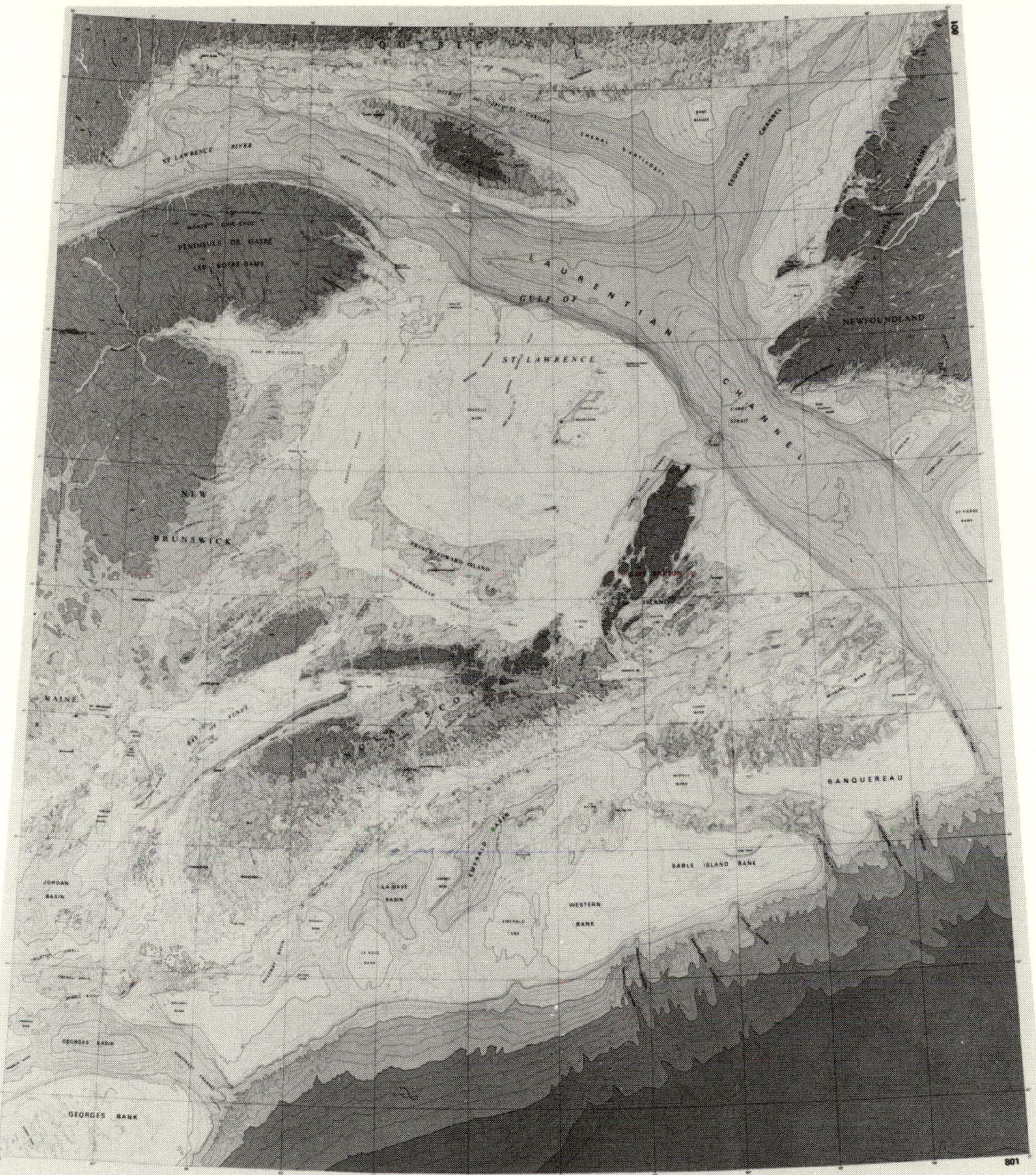

Fig. 8-2. Showing how the large North American ice sheets overlapped onto the continental shelves (black portions). Modified from Dott and Batten, *Evolution of the Earth*, (New York: McGraw-Hill Book Co., 1971).

Fig. 8-3. The deep trough extending into the Gulf of St. Lawrence and the irregular topography produced by the glaciation off southeastern Canada. From Canadian Hydrographic surveys. Contours by E. Uchupi.

though most are now greatly depleted, principally by Soviet trawler activity.

I originally interpreted these banks off glaciated coasts as moraines built up at the margin of the ice caps (Shepard, 1948, p. 151). As they have become much better known, due to various drillings and the collection of numerous samples as well as by exploration with sound-penetrating devices, it appears that this interpretation was only partly correct. Many of the shoals were built by streams emerging from the glaciers and building fans across the exposed outer continental shelves. Deltas were built into the adjacent seas. The same plains were developed on the continents in many places at the margins of the glaciers. These are generally called *outwash plains.* In fact, the relatively high northern portion of Cape Cod was due to stream deposition that occurred in be-

Fig. 8-4. The Grand Banks and the glaciated coast of Newfoundland. The various banks are a combination of glacial moraines and outwash deposits from the streams flowing out of the glaciers onto the emerged continental shelves. From Canadian Hydrographic Service. Contours by E. Uchupi.

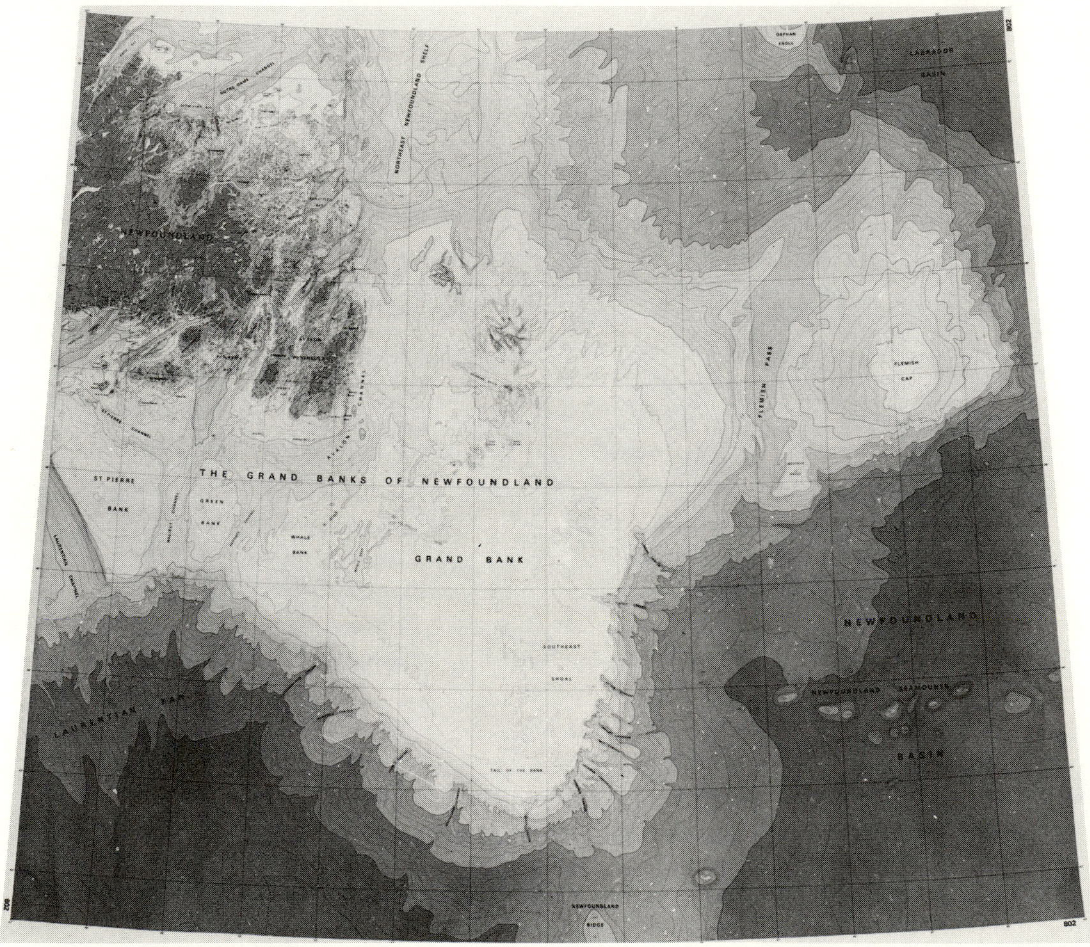

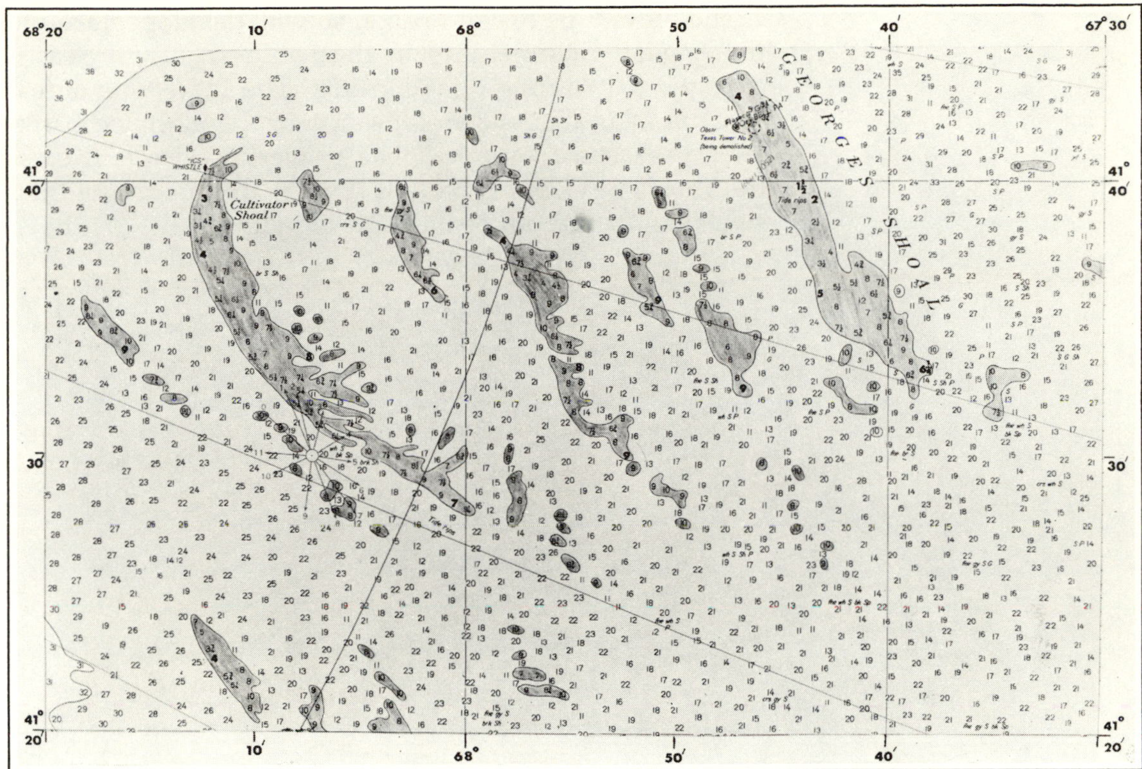

Fig. 8-5. The shoals on Georges Bank elongated by the tides sweeping across the bank. Depths are in fathoms. From U.S. Coast and Geodetic Survey charts.

tween two separate ice lobes. In some places, real glacial moraines have been found underlying these banks, notably off Nova Scotia (Stanley and Cok, 1967).

Shelves with Elongate Sand Ridges. Many continental shelves have long sub-parallel sand ridges rising a few fathoms above the general level. There appear to be at least two types of ridge. One is roughly parallel to the coast (Fig. 8-6), and the other extends at right angles to the margin but converges into embayments, like the English Channel or the southern North Sea (Fig. 8-7). Some disagreement among marine geologists (Sanders, 1963 and Swift, 1969) concerns the origin of the parallel ridges. They seem to resemble barrier islands along the present coasts, so they are generally interpreted as old barriers that were drowned by the rise in sea level due to melting of the last great ice caps. Objections to this idea stem from the discovery that some of the ridges near the shore migrate during storms, suggesting recent origin. Probably the present shape of the sand ridges is the result of storm modification, but their original development could easily be explained by drowning of barriers.

The other type of sand ridge, so well developed around the English Channel and in the southern North Sea, is clearly due to strong tidal currents. The same type of feature exists as a recent modification of the glaciated banks off northeastern America. The powerful tides that cross the shoals of Georges Bank (Fig. 8-5) are due to the tides sweeping in and out of the Gulf of Maine.

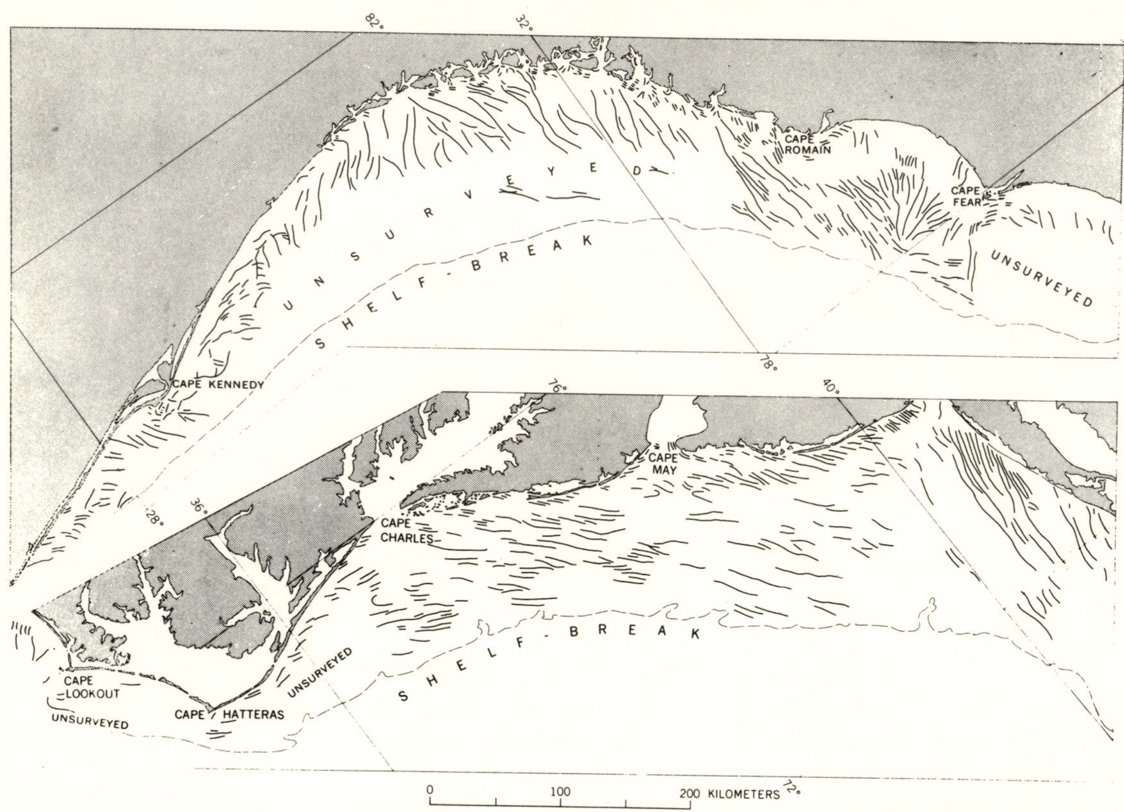

Fig. 8-6. Submerged sand ridges on the continental shelf from Long Island to Florida. From E. Uchupi, *U.S. Geological Survey Professional Paper 529 C*, 1968.

Some sand ridges are found in shallow water relatively near the coasts. A well-known example is Nantucket Shoals, near Nantucket Island. These are very unstable, being greatly shifted by storm waves. I was made aware of this after a small hurricane had hit the area in 1945. I was navigating a boat around the shoals, keeping in what the chart showed was adequate water depth. Suddenly we ran aground. The shoal had shifted. Grounding on a sand bar is better than hitting a rock. You can generally work your way off by shifting weights from side to side and kedging from the stern with the anchor, or more easily if you are fortunate enough to have another boat to tow you into deeper water. If at low tide, keep out your anchor

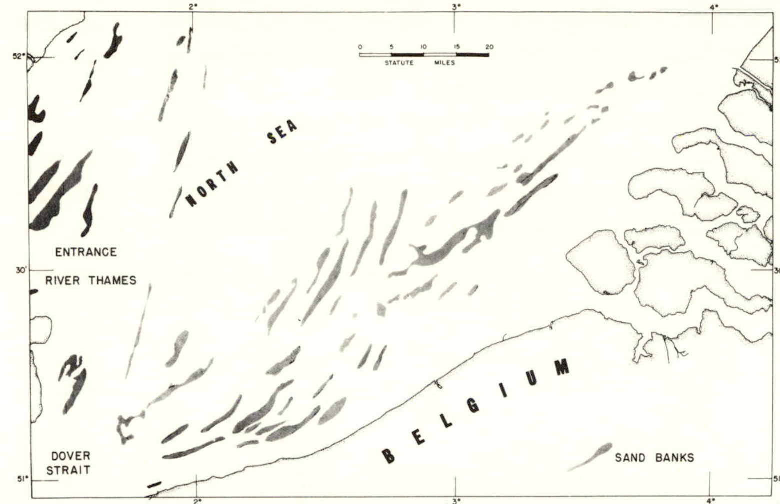

Fig. 8-7. The elongate irregular sand banks that characterize the southern end of the North Sea, formed primarily by tides.

with a taut line and the rising tide will free your boat.

Shelves Off Large Deltas. As in the case of the elongate sand ridges, the shelves adjoining deltas appear to be of two principal types. One has a shelf edge that bends seaward roughly parallel to the margin of the delta. This is exemplified by the Niger of West Africa (Fig. 8-8) and the Nile of Egypt. Clearly, the continental shelf has been built outward as these deltas advanced. The other type has a delta that grows across the continental shelf and a shelf margin that does not bulge appreciably seaward. The Mississippi is an example of the latter. In fact, two of the

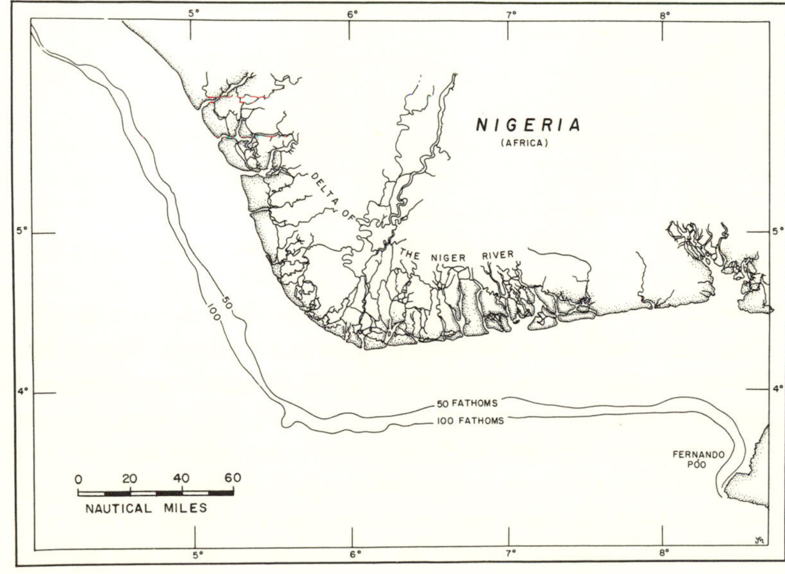

Fig. 8-8. The Niger Delta and the conformation of the shelf edge to the margin of the delta showing outward shelf growth at the same rate as advance of the delta.

passes of the Birdfoot Delta (Fig. 8-9) have reached the shelf edge, and deposition is now occuring on the gently inclined continental slope.

The shelves off both types of delta are shallower at the margins than average shelves. There are also some very flat terraces on these shelves that probably represent deltas that were drowned by the postglacial rise in sea level. Another feature of delta shelves is the oval hills. Some of

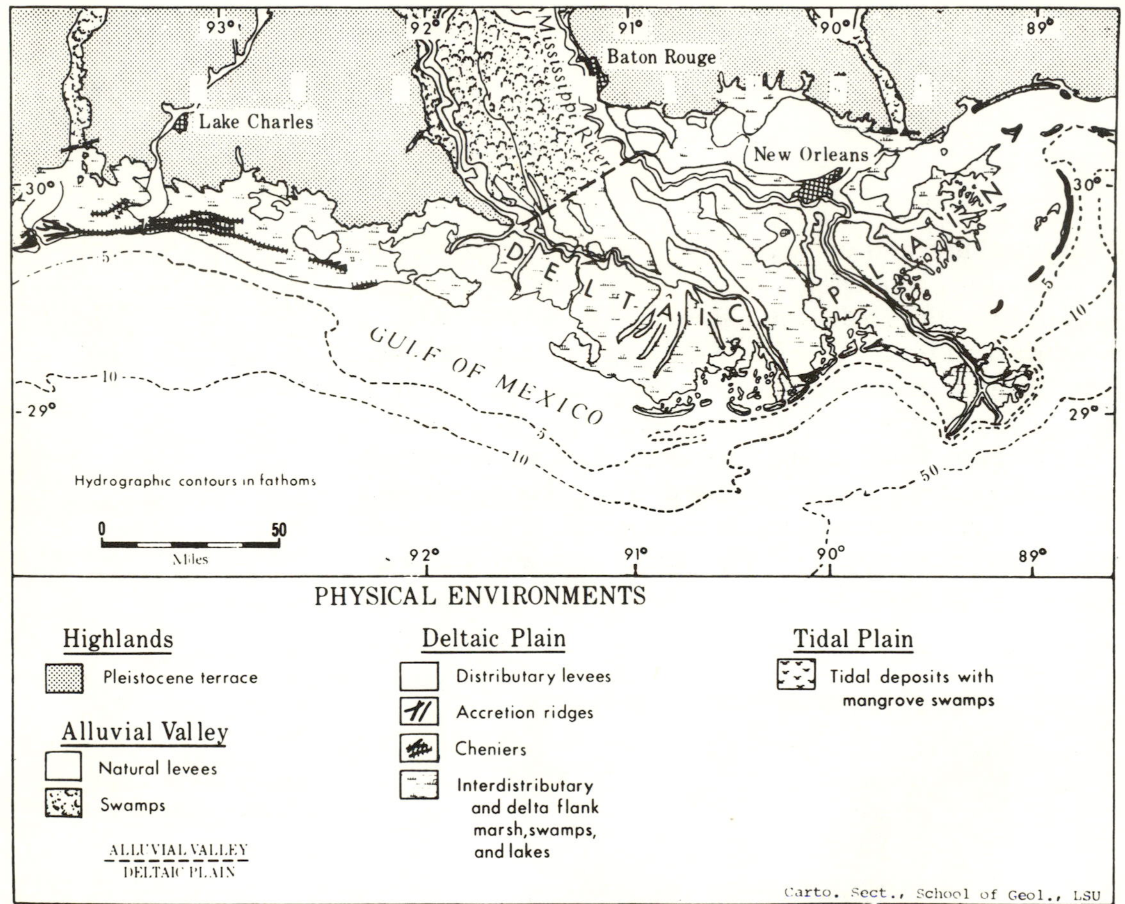

Fig. 8-9. The Mississippi Delta showing how the Birdfoot Delta has been built across the shelf in contrast to the Niger Delta shown in Fig. 8-8.

these come to the surface, as off the Mississippi where they are called *mud lumps* (Morgan et al., 1963) (Fig. 8-10). These small islands may pop up in a few hours, largely due to pressure of the advancing delta squeezing up underlying mud layers. The larger hills are called *diapirs* (Fig. 8-11). These are very common in the Gulf of Mexico (Fig. 8-12). The hills may also be large mud lumps, or may be *salt domes* where a buried salt layer has squeezed up the

Fig. 8-10. A mud lump island off the Mississippi Delta that rose rapidly and is being destroyed by waves. Other mud lumps in the distance.

overlying sediment. Oil is often found around these diapirs, particularly off Texas and Louisiana (Lehner, 1969).

Shelves with Coral Reefs. Coral reefs are discussed in the next chapter, but their presence on the continental shelves give us a reason to discuss them here. The shelves have zones very close to sea level and have many small islands rising a few feet above the wash of normal waves. The Great Barrier Reef off northeastern Australia is by far the largest shelf of this kind. At the south end of this reef is a wide channel extending inside most of the reefs, and the

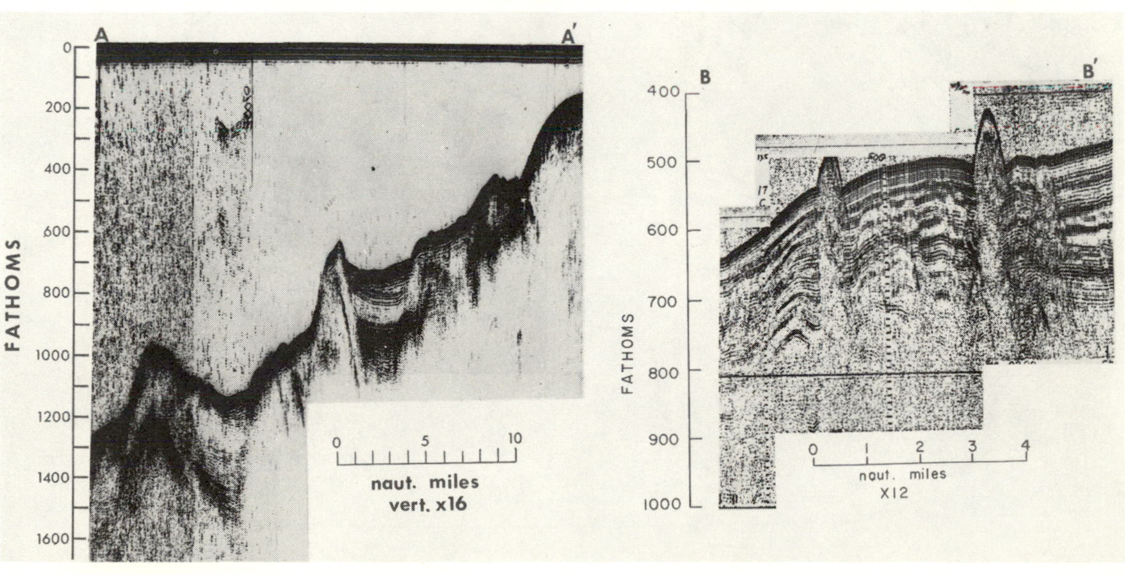

Fig. 8-11. Diapiric intrusions of mud on the slope off the Magdalena Delta, Colombia. From seismic profiles made on the R/V *Thomas Washington*.

outer shelf is relatively deep but has a drowned reef rising above its margin. Farther north, the channel is missing and the reefs extend in patches across the entire shelf.

Shelves Bordered by Rocky Banks and Islands. Many of the continental shelves have rocky banks along their outer margins (Fig. 8-13). These banks are remnants of old rock islands, and in some places portions of the islands still exist, as for example, the granite Farallones off San Francisco. The largest island of this type is Spitsbergen, north of the glacially excavated Barents Sea shelf of Norway and the Soviet Union (Fig. 8-14). The smaller Pribilof Islands rise above the outer portion of the broad Bering Sea shelf.

Rocky banks occur on the wide shelves, like the Bering Sea and Barents Sea, and on narrow shelves, as off Southern California. Apparently, the smoother sediment-covered portions of the shelves near shore have been built up by sedimentary deposits, partially filling basins and channels inside the outer rock ridges.

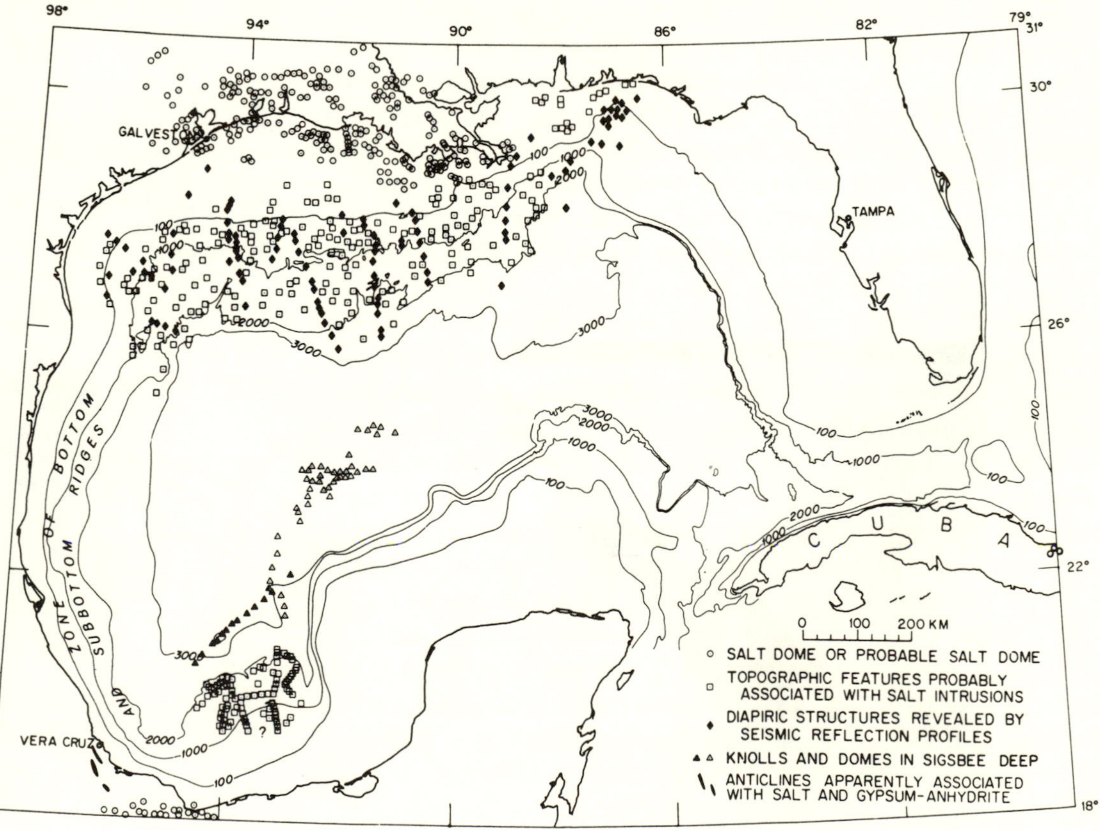

Fig. 8-12. Diapirs and salt domes on the continental shelf and slope off the Gulf Coast of the United States and in the deep central portion of the Gulf of Mexico. Note relation to oil-producing salt domes on land. From E. Uchupi, *Transactions Gulf Coast Geological Association,* 1967.

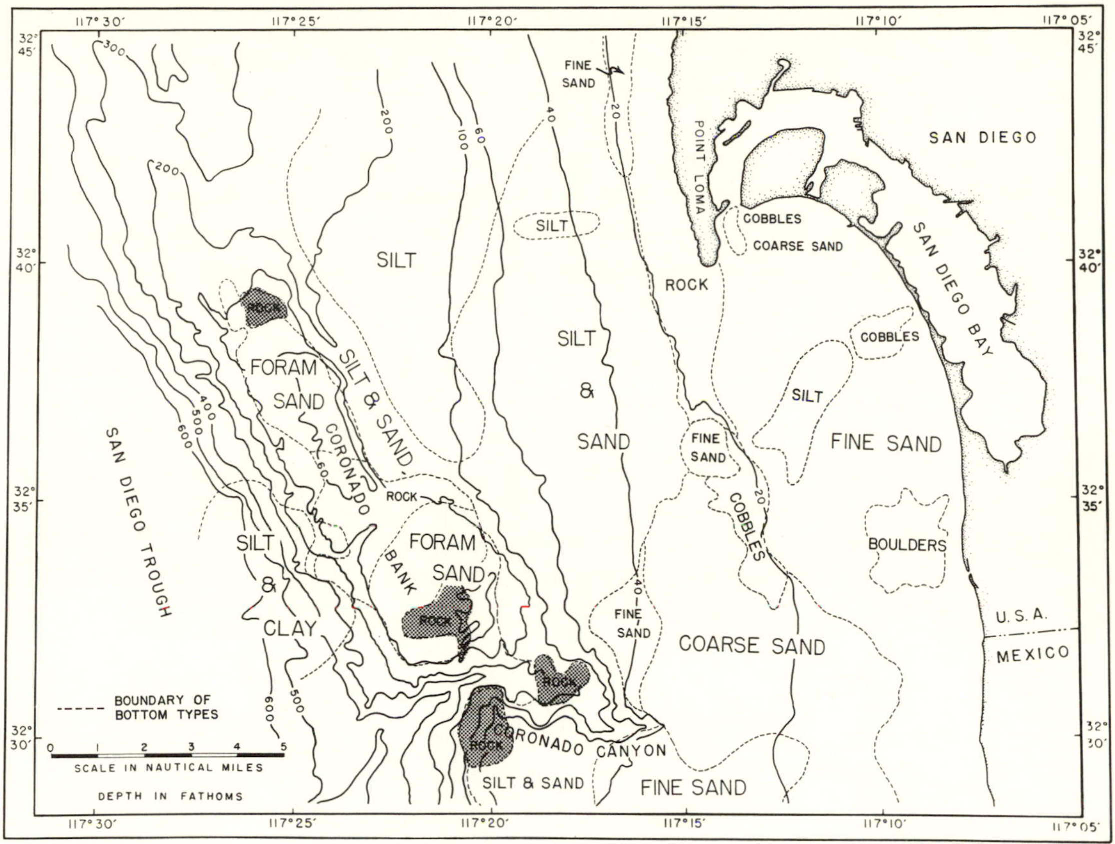

Fig. 8-13. A rock bank at the outer edge of the shelf off San Diego, California. Shaded area is rock. From K. O. Emery and others, *Journal of Geology,* 1952.

Shelves Related to Plate Tectonics. Assuming that the sea-floor spreading (or plate tectonics) hypothesis is correct, it is evident that the character of many shelves is closely related to these giant shifts of the earth's crust. Thus, the subduction of the crust under the advancing Andean coast of South America should have left little opportunity for a continental shelf to develop. Examination of the coastal charts shows that this assumption is borne out. Very narrow shelves are found along almost the entire west coast of South America. However, shelves along the Atlantic Coasts are relatively broad, for example, the east coast of the United States, the northeast coast of South America, and Argentina. The narrow shelf off northern Spain and southwestern France can be explained by a supposedly recent opening of the Bay of Biscay, which is a popular idea among plate tectonics geoscientists (Blackett et al., 1965).

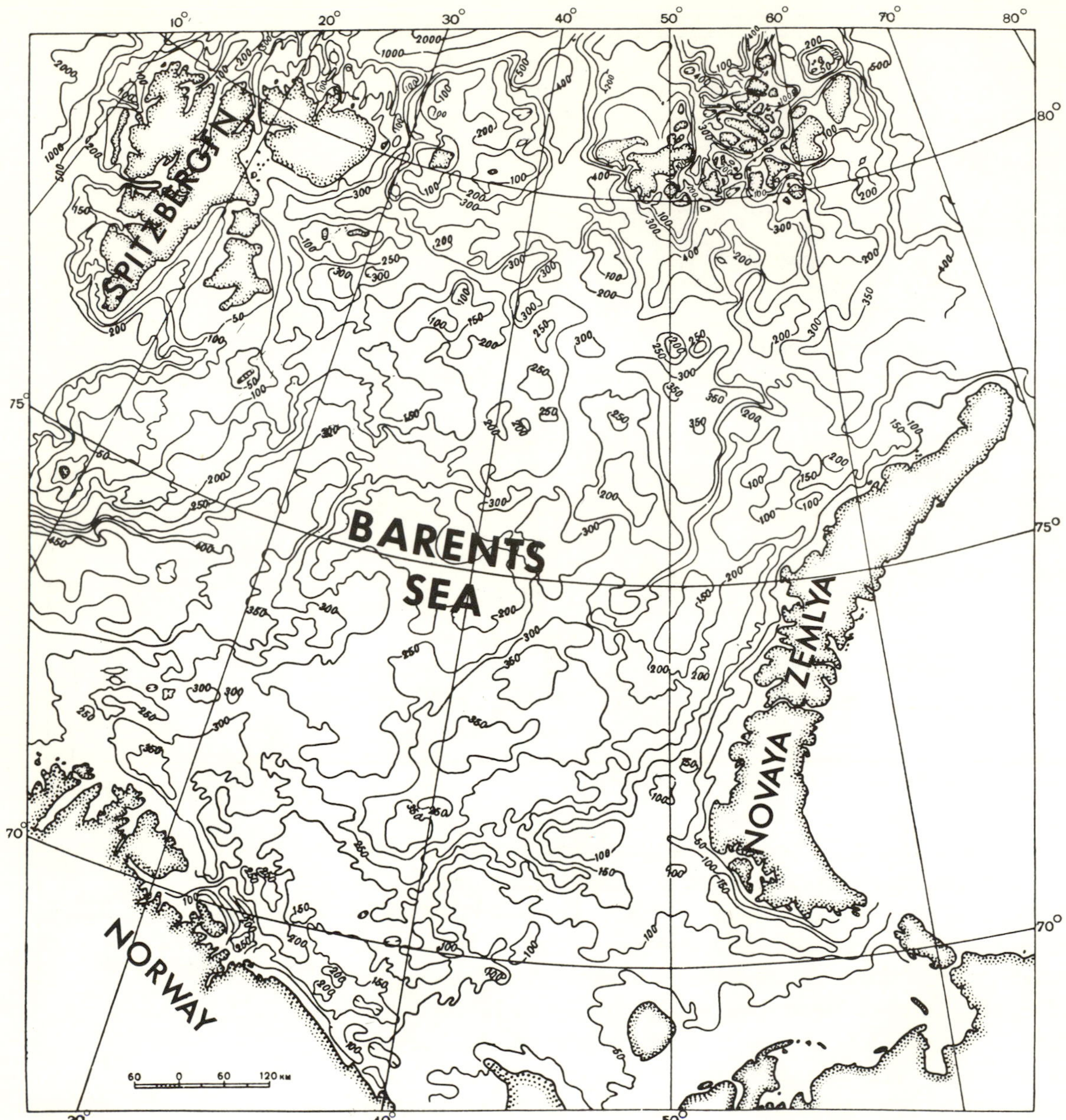

Fig. 8-14. Glacial erosion topography in the Barents Sea shelf, north of Norway and the Soviet Union. Depths in fathoms. From M. V. Klenova, Academy of Sciences, USSR.

Where plate margins have been pushed counter to each other as, according to theory, the Caribbean-West Indian plate has pushed past the South American plate, one could expect narrow continental shelves, and this is borne out along most of the north coast of South America.

Continental Shelf Sediments

As indicated in the introductory chapter, the sediments on the continental shelves show a surprising absence of a gradation from coarse near shore to fine on the outer margins. A typical sediment map is shown in Fig. 8-15. This reminds one of a crazy quilt. At first glance it seems to have no relationship to the activity of the waves and currents. However, the arrangement of sediment bands and patches becomes far more understandable when we make use of what we know about the late history of sea-level rise. During the lowered sea level of the Ice Age, when the seas had exposed most of the shelves, the coarse sedi-

Fig. 8-15. Patchy distribution of sediments in the North Sea and surrounding seas. From O. Prattje, *Handbuch der Seefischerei Nordeuropas*, 1949.

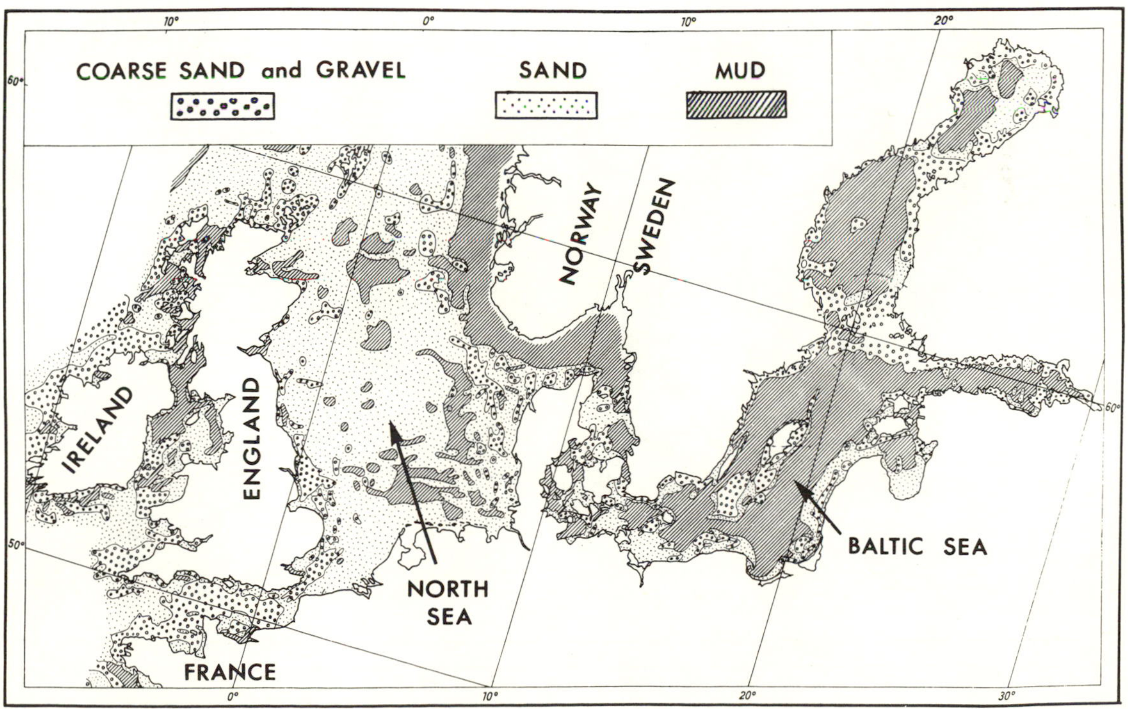

ments could be carried by the waves and currents over extensive portions of the outer shelves. Also, the rivers in general were flowing faster, partly from increased rainfall and stormier conditions that existed in many areas and partly due to steeper gradients. Hence, coarse sediments were often deposited at the shelf margin.

As the sea level rose, drowning land valleys so that streams flowed into bays, the ocean currents effectively prevented a cover of the coarse sediments on the outer shelves. In fact, the extensive collection of coarse sediments from the outer shelves contains many fossils that have been dated as from the same time as the glacially

Fig. 8-16. Showing the coarsening of sediments on the outer shelf off the Orinoco Delta. From Tj. H. van Andel, *Jour. of Sedimentary Petrology*, 1967.

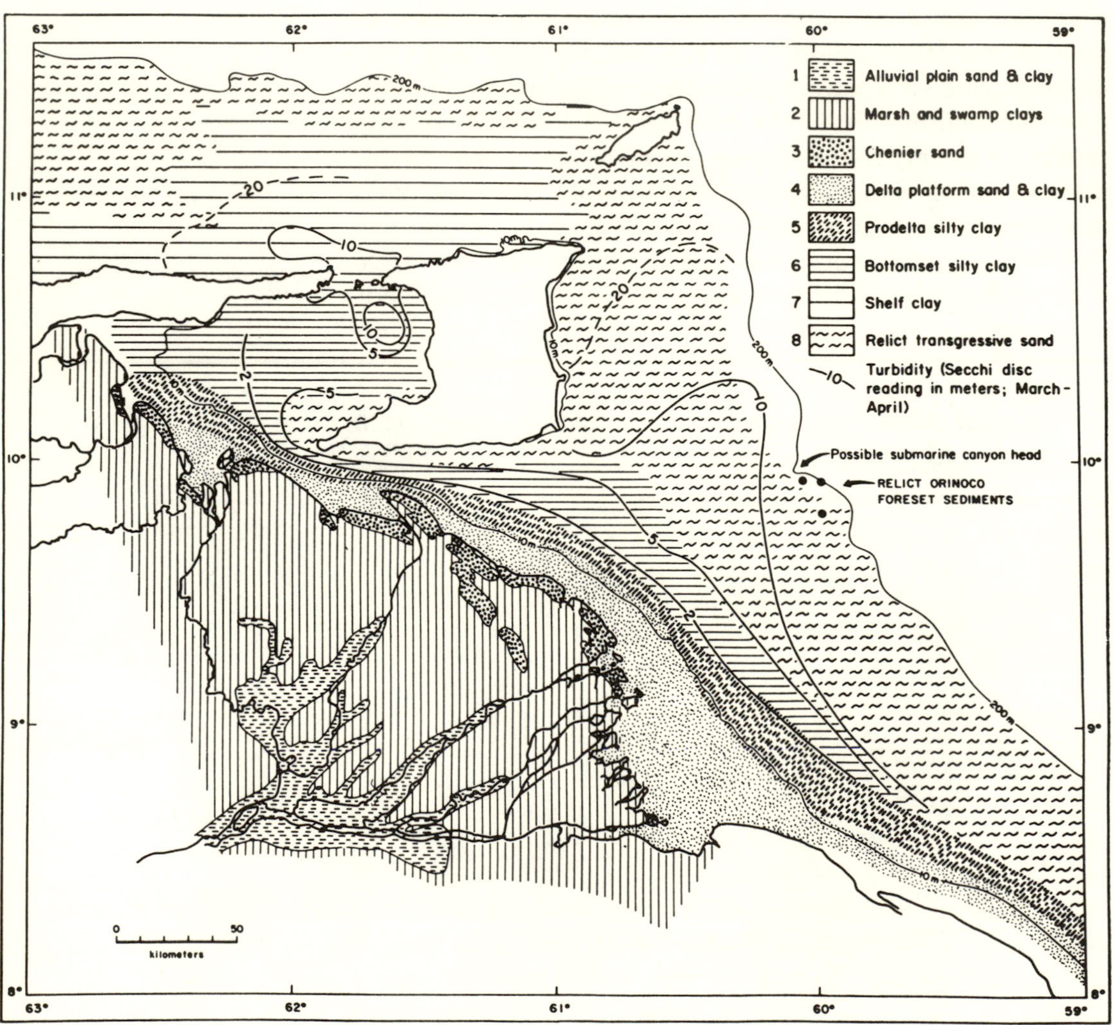

lowered sea levels. The group working under Kenneth Emery's supervision at Woods Hole Oceanographic Institution found many bones of land animals, including mammoths, in the superficial sediments of the East Coast shelves (Emery, 1965).

How about the sediments near shore? Here again we find some explanations of the sediment distribution. Muds often form a blanket over the inner shelf off large river mouths (Fig. 8-16). In many places, the outer shelf off deltas is sand covered. Mud bottom is also found in many shelf depressions, whereas hills and ridges usually have coarse sediment or even rock. The explanation of the mud areas off large rivers is that so much fine sediment is introduced that the currents are unable to keep it moving to deeper water. Currents on the outside have less sediment to deal with, so sand that was deposited during lower sea levels may have remained uncovered. Coarse sediment on hills and ridges is also the remnant of earlier sea level conditions, and present-day currents are usually stronger on these high areas of the shelf.

Where currents are concentrated in funnel-shaped embayments, the shelf depressions may lack the usual mud sediments. An example is the elongate depressions that extend up the English Channel. Rock has been dredged from many of them and sand is common.

The sea floor of glaciated shelves shows considerable variation. Mud bottom is common in the troughs and basins, but cores usually show much sand and many rock fragments mixed in with the mud. The coarse sediment is often ice-rafted, coming from the melting glaciers or from the spring breakup of river ice after the glaciers had retreated.

Shelf Origin

We have already discussed the origin of some of the types of continental shelf, but must now consider the shelves as a whole. Why do these shallow platforms flank most of the coasts of the world and terminate seaward with a definite increase in slope? The old explanation was simplicity itself. The waves cut back the coasts, making wave-cut benches, and the sediment cut from the retreating land is deposited seaward of the wave-cut bench as a wave-built terrace (Fig. 8-17A). In preliminary investigations of marine geology, many objections developed to this idea. We discovered rock bottom on the outer portions of shelves in numerous areas and dredged rock from the slopes beyond, particularly where the slopes were cut by canyons. According to the old hypothesis, the wave-cut and wave-built terraces should have a general relationship to the size of the storm waves along the various open coasts, but we actually found that the shelf margins were shoaler off the exposed stormy coasts than off those with a predominance of offshore winds and hence smaller waves. Furthermore, the continental slopes showed little relationship to those off submerged delta fronts. For these and other reasons the simple explanation did not hold up.

We now have numerous seismic profiles that show us the underlying structure of the outer shelf. Many of these new profiles do show that a sedimentary embankment has built out the shelves (Fig. 8-17B). However, the built terraces are almost always complicated by erosion, faulting, and slumping (Fig. 8-17C). In many other places, the shelf edge has a rock core and no sediment terrace on the outside (Fig. 8-17D). Elsewhere, the slope is cut by faults (Fig. 8-17E) and landslides are very common along the slopes (Fig. 8-17F). Some slopes are underlain by a series of folded layers (Fig. 8-17G). Thus the shelf may or may not be related to outbuilding of sediment. Furthermore, the progradation may be in the form of a delta, as off the Nile, the Niger, and the Mississippi.

Without doubt, glacial sea-level lowering has had an important influence on shelf development. When the sea was lower, the waves cut benches at or just below the low levels. Many of the platforms on the outer shelves have been found to represent glacial stage cutting. Also, deltas were built out during the low sea-level stages, and some were left without sediment cover when the huge glaciers melted. Others had superimposed deltas built during or after the rising sea level.

So again we see an emerging picture which indicates that the old hypothesis was too simple, and a combination of causes makes a better explanation (see also Hedberg, 1970, and Emery and Uchupi, 1972).

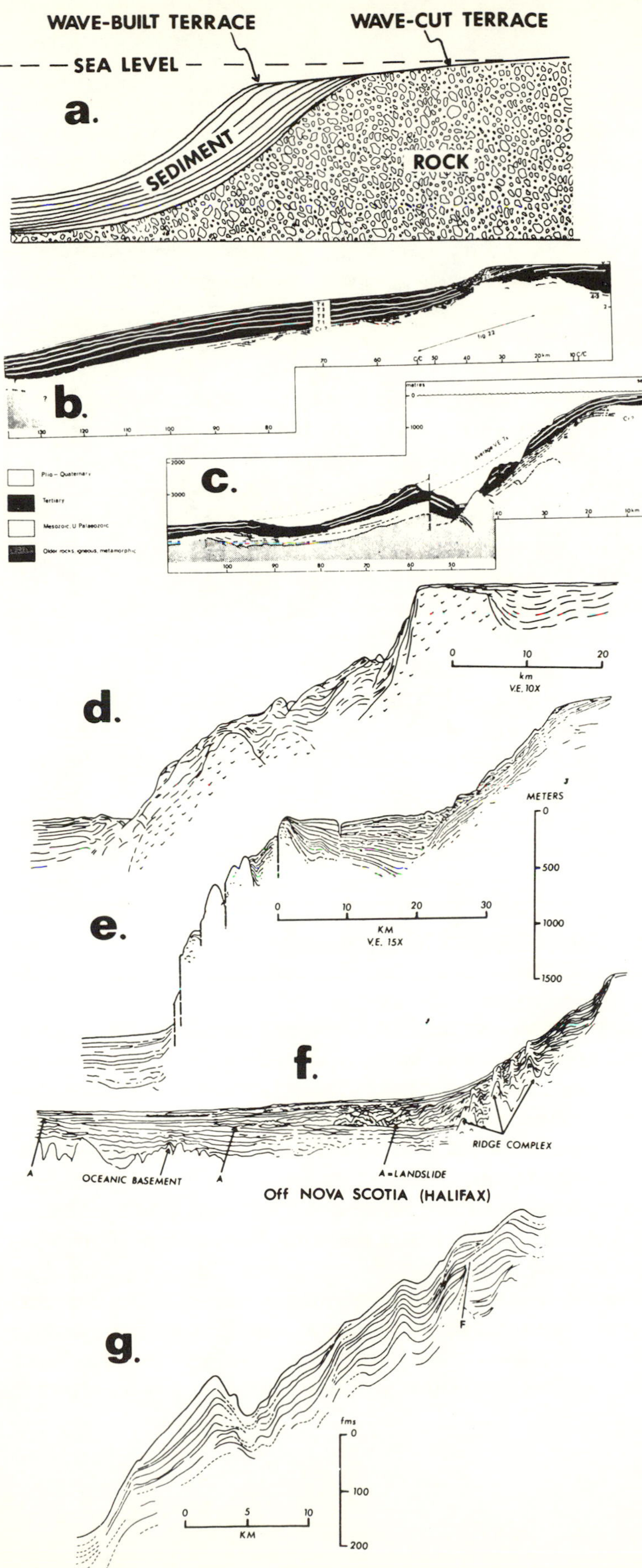

Fig. 8-17. Various types of continental slopes shown by seismic profiles: (A) theoretical wave-cut, wave-built terrace, rarely found in nature; (B) prograded slope of southwestern France; (C) prograded slope cut by slumping and faulting; (D) granite ridge on outer shelf off San Francisco, and slope cut by extensive slumping. Note sediments on inner shelf; (E) extensively faulted continental slopes off Northern California; (F) a large low-angle landslide on the slope off Nova Scotia; (G) folds probably due to slumping off southern Portugal.

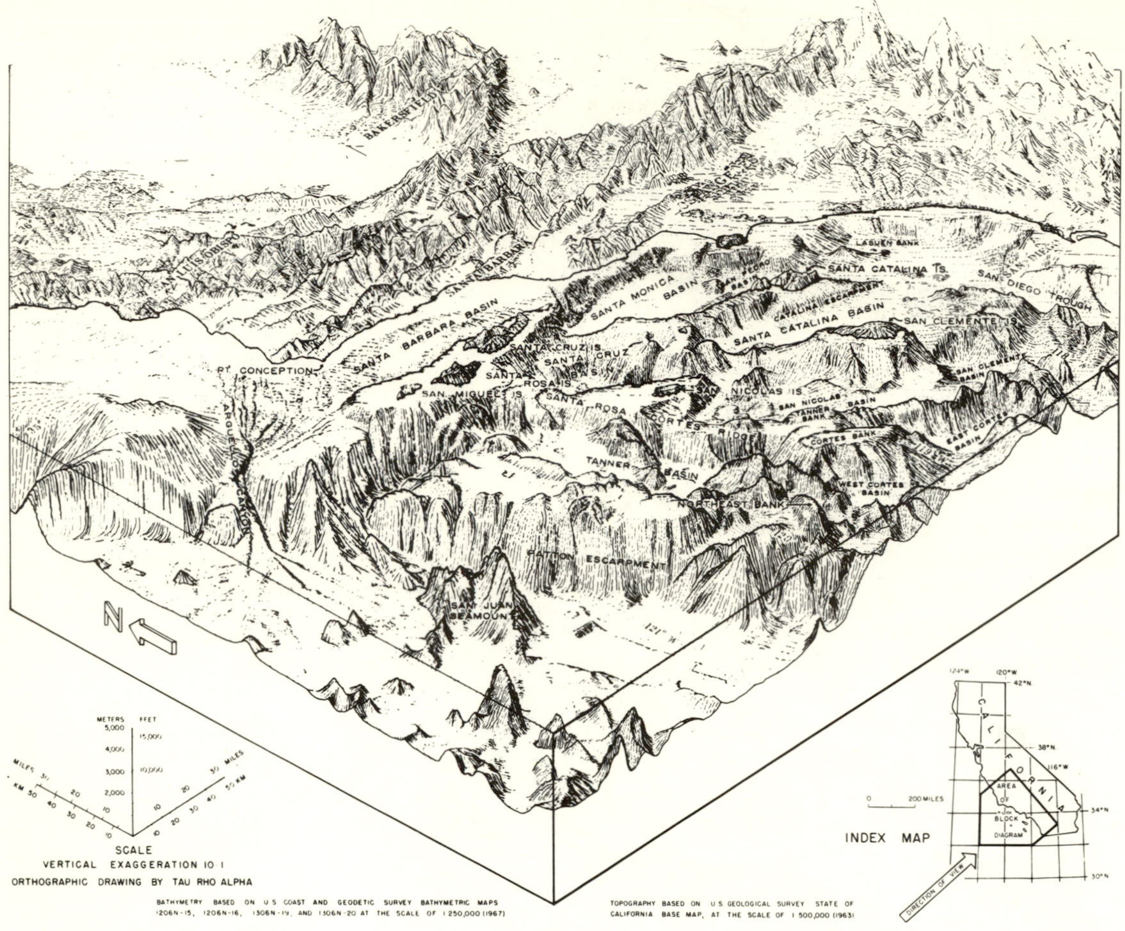

Fig. 8-18. The continental borderland off Southern California and land area (upper portion). Block diagram by T. R. Alpha, U.S. Geological Survey.

Continental Borderlands

In some areas, instead of finding a simple shallow continental shelf bordered by a slope leading to the deep ocean, we have discovered either a deep terrace beyond the continental shelf or a series of basins and ridges extending for many miles beyond a narrow shelf. These, called *continental borderlands*, are found off many coasts but the two best known are located off Southern California and southeastern United States. The California Borderland (Fig. 8-18) has several high islands, notably Catalina and San Clemente Island. There are also a series of basins and troughs with depths to about a mile (1.6 km). Relatively flat banks occur on the borderland, and Cortes Bank has a rock reef that is so close to the surface that it is a danger to navigation. The California continental borderland is the result of comparatively recent faulting that dropped the basins and troughs below the islands, banks, and shelves. A steep fault scarp forms the outer margin of the borderland leading to the deep ocean floor. Considerable sedimentation has occurred in the basins and troughs, but currents have swept the banks relatively bare so that rock outcrops in many places on them. The banks probably contain abundant oil deposits.

The other type of continental borderland is located south of Cape Hatteras, terminating near the Straits of Florida (Fig. 8-19). Here there is a deep terrace, called Blake Plateau, where the water depth is mostly between 400 and 600 fathoms. This plateau, actually a terrace, has a different origin. The Gulf Stream sweeps out almost the entire surface. In 1969 Jacques Piccard and a group of aquanauts made a voyage in the deep-diving *Ben Franklin* all along this terrace, and other scientists have explored the bottom, both by deep dives and by sampling from

Fig. 8-19. The deep Blake Plateau that borders the narrow shelf south of Cape Hatteras and adjoins the Bahama Islands on the south.

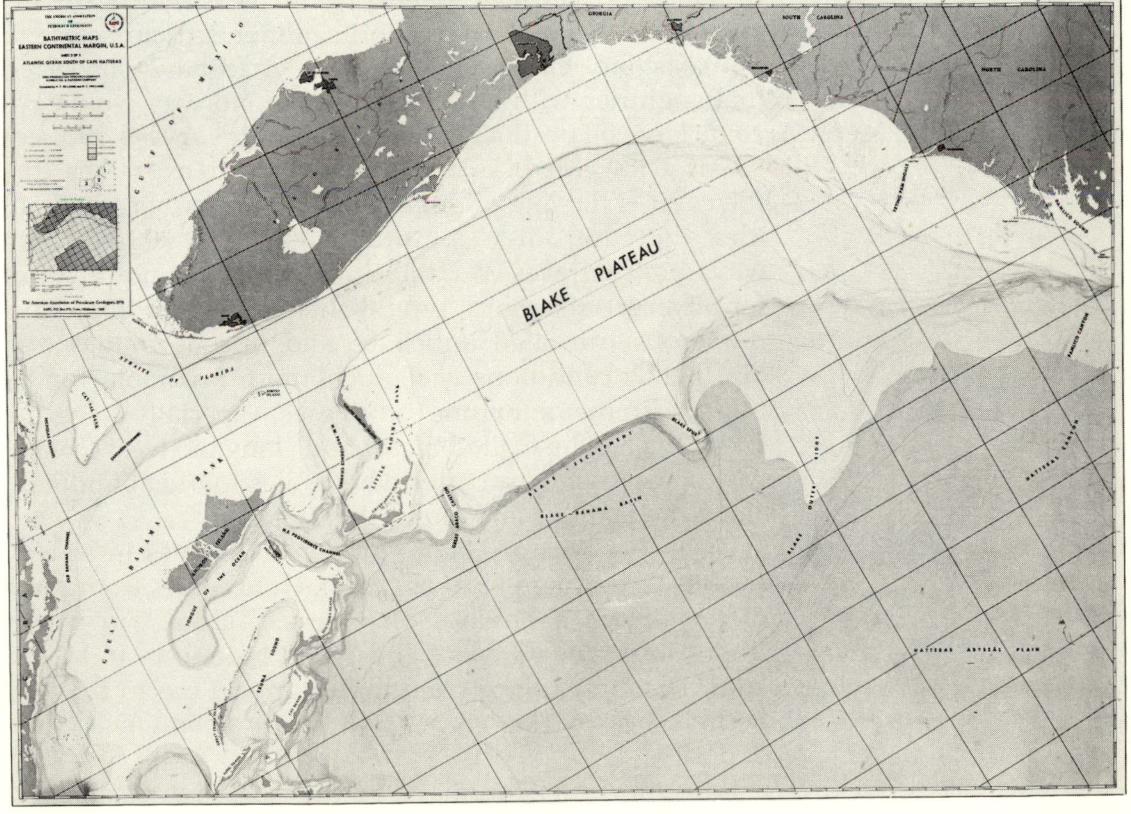

ships. The platform is essentially a rock surface with little sediment. It seems probable that the Gulf Stream or its predecessor has prevented deposition of sediments for many millions of years. Probably the platform was once near sea level and has sunk to its present level. Meanwhile, sediments have been building up the sinking continental shelves in the area north of Cape Hatteras. The outer margin of Blake Plateau is very steep. Probably faulting has occurred, but the borings show that the steep slopes may be due to building up of a coral reef as the area sank.

Economic Resources of the Continental Shelves and Borderlands

As has already been mentioned, there are vast areas on the continental shelves where petroleum and gas are being recovered. It is surprising how often contact between old rock masses (which are rarely productive of hydrocarbons) and relatively young rocks and sediments occurs close to the coastline. Geologists seem never to have suspected that countries with ancient rocks like Norway would become large oil producers, but the seismic surveys off these coasts with old rocks have shown in many places that the coast is the line of demarcation. Furthermore, the old idea that the outer continental shelves were simply wave-built terraces of monoclinal dip could not account for oil structures. Since typical shelf seismic profiles have numerous truncated anticlines and unconformities (see Fig. 8-17), conditions are much more favorable for oil traps. Also the numerous salt domes, especially along the Gulf Coast of the United States and along the west coast of Africa, represent structures where oil is found abundantly along the margins and above the domes.

Placer deposits are being mined on shelves in several regions (Emery and Noakes, 1968). Gold comes from some of the drowned beaches off Nome, Alaska. Tin comes from submerged alluvial deposits off Indonesia and Thailand. Diamonds have been mined from time to time off South Africa. These are also from submerged river deposits.

Phosphorite is found on many shelves, notably off

southern California in the San Diego area (Dietz et al., 1942). Several companies are considering mining these nodules. One thing that has held up the operation was the dumping of large numbers of explosives in the phosphorite areas after World War II. Iron is mined in small quantities near Conception Bay, northeast Newfoundland.

It seems likely that more exploration will reveal other shelf deposits that are sufficiently economic in quantity to start operations. Finally, there is likely to be a large amount of sand quarried from the extensive deposits of sand on the continental shelves off areas where beaches are rapidly disappearing or where road material is desperately needed.

References

Blackett, P. M. S., E. Bullard, and S. K. Runcorn, 1965. *A Symposium on Continental Drift.* Philosophical Trans. No. 1088, The Royal Society, London, p. 49.

Dietz, R. S., K. O. Emery, and F. P. Shepard, 1942. "Phosphorite deposits on the sea floor off Southern California." *Geol. Soc. Amer. Bull.,* v. 52, pp. 815-848.

Donovan, D. T., ed., 1968. *Geology of Shelf Seas, Proceedings of the 14th Inter-University Geological Congress.* Oliver & Boyd, London, 160 pp.

Dott, R. H., Jr. and R. L. Batten, 1971. *Evolution of the Earth.* McGraw-Hill Book Co., N.Y., 649 pp.

Emery, K. O., 1965. "Geology of the continental margin off eastern United States." In *Submarine Geology and Geophysics,* W. F. Whittard and R. Bradshaw, eds. Butterworths, London, pp. 1-20.

Emery, K. O., 1967. "Geological Aspects of Sea-floor Sovereignty." Ch. 9 in *The Law of the Sea,* L.M. Alexander, ed., Ohio Univ. Press, pp. 139-159.

Emery, K. O., 1970. "An oceanographer's view of the law of the sea." Woods Hole Oceanographic Institution, Contrib. No. 2360.

Emery, K. O., and E. Uchupi, 1972. *Western North Atlantic: Topography, Rocks, Structure, Water, Life and Sediments.* Amer. Assoc. Petrol. Geol., Tulsa, Okla. Memoir 17, 532 pp.

Emery, K. O., W. S. Butcher, H. R. Gould, and F. P. Shepard, 1952. "Submarine geology off San Diego, California." *Jour. Geol.*, v. 60, no. 6, pp. 511-548.

Emery, K. O., and L. C. Noakes, 1968. "Economic placer deposits of the continental shelf." *Tech. Bull. ECAFE*, v. 1, pp. 95-111.

Emery, K. O., E. Uchupi, J. D. Phillips, C. O. Bowin, E. T. Bunce, and S. T. Knott, 1970. "Continental rise off eastern North America." *Amer. Assoc. Petrol. Geol. Bull.*, v. 54, no. 1, pp. 44-108.

Gregory, J. W., 1930. "The earthquake south of Newfoundland and submarine valleys." *Nature*, v. 124.

Hedberg, H. D., 1970. "Continental margins from viewpoints of the petroleum geologist." *Amer. Assoc. Petrol. Geol. Bull.*, v. 54, no. 1, pp. 3-43.

Holtedahl, Olaf, 1940. "The submarine relief off the Norwegian coast." *Norske Videnskaps-Akad.*, Oslo, 43 pp.

Johnson, D. W., 1925. *New England—Acadian Shoreline.* Wiley, N.Y., 608 pp.

Keith, A., 1930. "The Grand Banks earthquake." *Seismol. Soc. Amer., Eastern Sec. Proc.*, Suppl.

Klenova, M. V., 1940. "Sediments of the Barents Sea." *Comt. Rend. (Doklady) Acad. Sci. URSS*, v. 26, no. 8, pp. 796-800.

Lehner, P., 1969. "Salt tectonics and Pleistocene stratigraphy on continental slope of northern Gulf of Mexico." *Amer. Assoc. Petrol. Geol. Bull.*, v. 53, no. 12, pp. 2431-2479.

Morgan, J. P., 1970. "Depositional processes and products in the deltaic environment." In *Deltaic Sedimentation, Modern and Ancient*, J. P. Morgan, ed., Soc. Econ. Paleontol. and Mineralog., spec. publ. no. 15, Tulsa, Okla., pp. 31-47.

Morgan, J. P., J. M. Coleman, and S. M. Gagliano, 1963. "Mudlumps at the mouth of South Pass, Mississippi River; sedimentology, paleontology, structure, origin and relation to deltaic processes." *Coastal Studies Inst., Louisiana State Univ.*, ser. 10, 190 pp.

Pratje, O., 1949. "Dodenbedeckung der nordeuropäischen Meer." *Handbuch der Seefischerei Nordeuropas*, v. 1, pt. 3, 23 pp.

Sanders, J. E., 1963, "North-south trending submarine ridge composed of coarse sand off False Cape, Virginia." Abs., *Amer. Assoc. Petrol. Geol. Bull.*, v. 46, no. 2, pp. 278.

Shepard, F. P., 1948. *Submarine Geology*. Harper & Bros., N.Y., p. 151.

Shepard, F. P., R. F. Dill, and B. C. Heezen, 1968. "Diapiric intrusions in foreset slope sediments off Magdalena Delta, Colombia." *Amer. Assoc. Petrol. Geol. Bull.*, v. 52, no. 11, pp. 2197-2207.

Stanley, D. J., and A. E. Cok, 1967. "Sediment transport by ice on the Nova Scotian Shelf." In *Ocean Sciences and Engineering of the Atlantic Shelf*. Trans. Nat. Symp. Mar. Technol. Soc., pp. 100-125.

Swift, D. J. P., 1969. "Outer shelf sedimentation: processes and products." In *The New Concepts of Continental Margin Sedimentation*, Lecture no. 5, Amer. Geol. Inst. Short Course Lecture Notes, Phila., Amer. Geol. Inst., Washington, DC, DS 5, 26 pp.

Uchupi, E., 1967. "Bathymetry of the Gulf of Mexico." *Trans. Gulf Coast Assn. Geol. Soc.*, 17th Ann. Meeting, pp. 161-172.

Uchupi, E., 1968. "Atlantic continental shelf and slope of the United States—Physiography." *U.S. Geol. Surv. Prof. Paper 529-C*, 30 pp.

van Andel, Tj. H., 1967. "The Orinoco Delta," *Jour. Sed. Petrology*, v. 37, no. 2, pp. 297-310.

Suggested Supplementary Reading

Burk, C. A., and C. L. Drake, eds., 1974. *The Geology of Continental Margins*. Springer-Verlag, N.Y., 1009 pp.

Curray, J. R., 1965. "Late Quaternary history, continental shelves of the United States." In *The Quaternary of the United States*, H. E. Wright and D. G. Frey, eds. Princeton Univ. Press, N.J., pt. 4, pp. 723-735.

Emery, K. O., 1969. "The Continental Shelves." *Sci. Amer.*, v. 221, no. 3, pp. 106-122.

Fairbridge, R. W., ed., 1966. *The Encyclopedia of Oceanography*. Reinhold Publ. Corp., N.Y., pp. 202-214.

Coral reef wonderlands

Chapter 9

If you have never snorkled over a healthy coral reef, you have missed one of life's most wonderful experiences. You don't need to be a scuba diver to see the beauty of the reefs and their schools of many-colored fish (Pl. 3). The corals grow up almost to the surface so you can skim over them and see some of the best of the underwater scenery. A glass-bottom boat really is a poor substitute for swimming with a face plate, because the boats cannot get over some of the best parts of the reefs and because vision is not as good through the thick boat bottom. Also, many of these boats operate around tourist centers where the water is somewhat murky, polluted by drainage.

Many of the earlier Cousteau movies were effective in showing the magnificent underwater panoramas of the reefs with their arches and overhanging fronds that allow good hiding places for some of the world's most beautiful fish, but you need to see these reefs for yourself to fully appreciate them. Remember that your chances of shark attack are considerably less than the danger of being hit by lightning.

What is a Coral Reef?

Coral reefs are by no means made entirely of corals. Some of them consist more of calcareous algae and lime-secreting shells than corals. However, the reefs that grow so luxuriously in tropical waters usually have a framework of branching corals that have become attached to a rocky bottom and have grown as a colony, both surfaceward and laterally, with the new corals branching to form shoots, while the old die and become fossilized underneath. The framework is very porous, but the skeletons of many animals and calcareous algae gradually fill the cavities between the branches. Boring animals, on the other hand, notably sea urchins and starfish, eat new holes into the reefs, so the reefs tend to remain very porous even after having been covered by thousands of years of new growth.

Fig. 9-1. Coral reef exposed by a very low tide, north coast of Moorea, Society Islands.

The reefs grow up to a level where they are just exposed at the lowest tides (Fig. 9-1). However, at the reef margin, a type of algal plant usually called *Lithothamnion* often grows slightly above sea level but never higher than can be reached by the spray of the breaking waves. Another common type of algae spreads over the bottom in lagoons inside the actively growing reef. This plant, called *Halimeda,* is an important constituent of the deposits in coral lagoons.

Conditions Necessary for Good Reef Development

Although one usually associates corals with shallow warm water, they actually grow at all depths and most temperatures. Deep-living cold-water corals, called *ahermatypic,* are usually not effective in building reefs. However, in recent years we have found some substantial deep ahermatypic reefs, notably along the outer rim of Blake Plateau and on the outer shelves off Norway. The *hermatypic* reef corals are confined to clear water less than about 400 feet (122 m) deep, with temperatures rarely getting below 68°F (20°C) and in water having an abundance of plankton to provide a food supply. Where rivers carry mud-laden water to the sea, the corals are scarce. Floods may kill an entire reef.

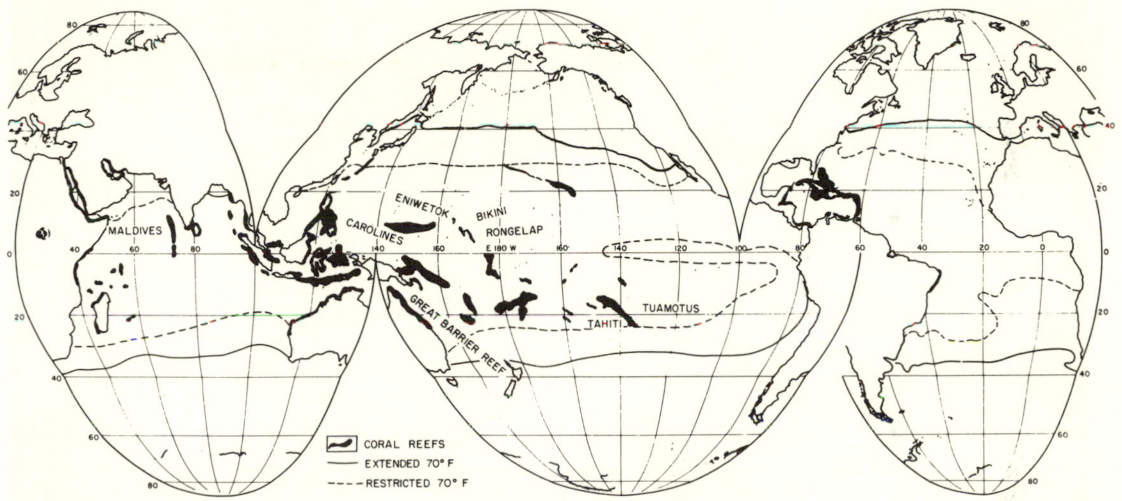

Fig. 9-2. World map of coral island areas.

The world map of coral reefs (Fig. 9-2) shows that the bulk of the reefs are on the western side of the oceans, and, because of the warmth necessary, in latitudes of less than about 23° North and South. Reefs are concentrated in western oceans because of the greater food supply that is brought to them by the west-blowing trade winds.

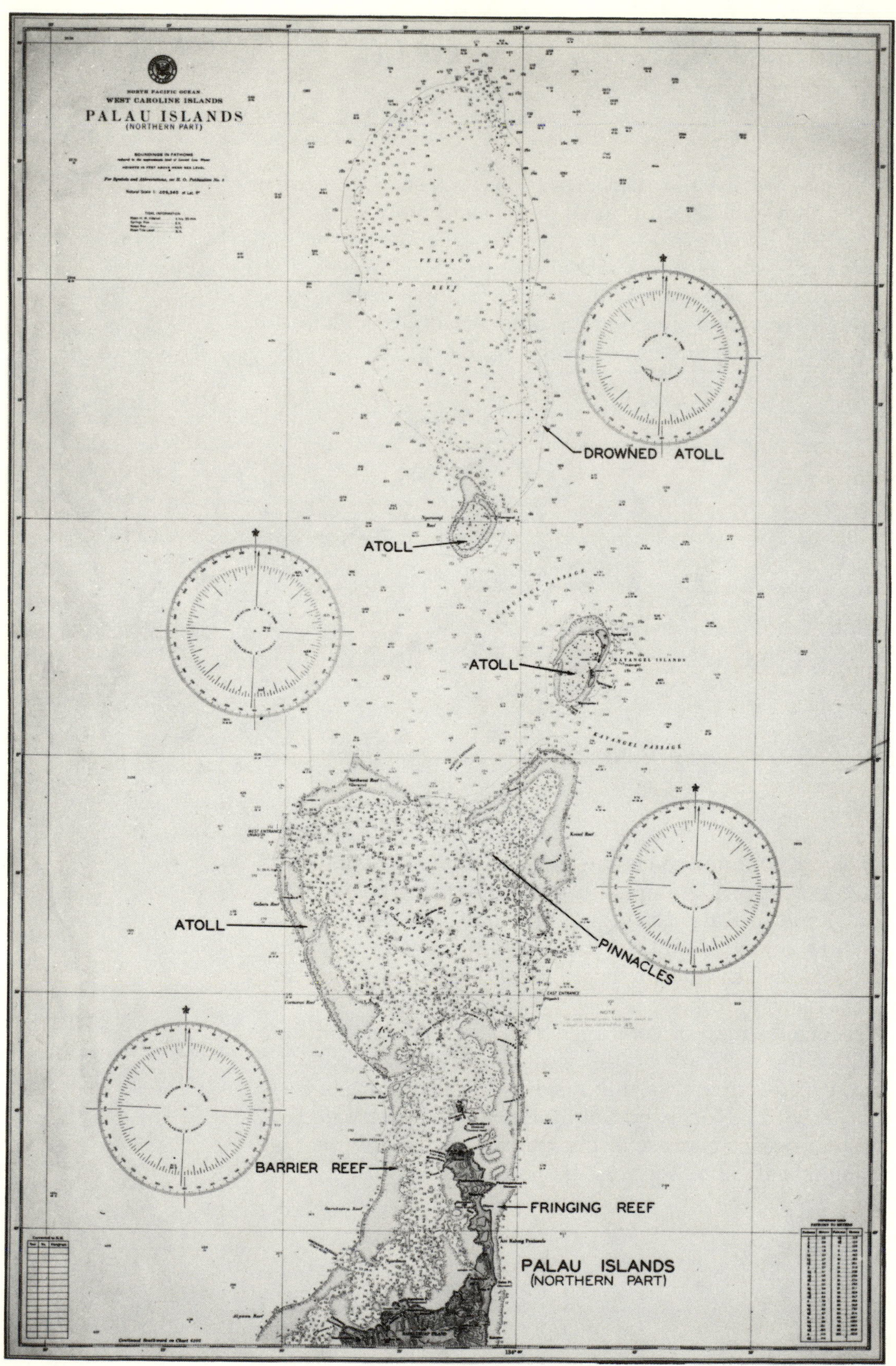
NORTH PACIFIC OCEAN
WEST CAROLINE ISLANDS
PALAU ISLANDS
(NORTHERN PART)
DROWNED ATOLL
ATOLL
ATOLL
ATOLL
PINNACLES
BARRIER REEF
FRINGING REEF
PALAU ISLANDS
(NORTHERN PART)

Reef Types and their Origin

There are four principal types of coral reef: fringing reefs, barrier reefs, atolls, and coral banks or patches (Fig. 9-3). It is now generally believed that a theory proposed by Charles Darwin in 1842 explains the evolution of three of these types. A reef starts to grow next to a rocky shore and extends gradually outward, forming a *fringing reef.* Then, due to sinking of the island or continental coast, the reef grows upward along the outer margin, leaving a lagoon on the inside and a *barrier reef* outside (Pl. 2, Fig. 3). Finally, if the sinking of an island carries the land entirely below sea level but the reef continues to grow to the surface, a ring of coral may still persist around the sunken land, forming an *atoll.* This is easily explained where the land mass was a volcano, as in Figure 9-4. The numerous atolls of the Southwest Pacific have supplied us with much evidence favoring this theory. After World War II, the U.S. Navy (cooperating with the U.S. Geological Survey) provided us the means of drilling deep in the Marshall Islands to test the hypothesis (Ladd et al., 1970). At Bikini the drilling went through 2500 feet (760 m) of shallow-water reef without getting to the underlying lava rock. But at Eniwetok, drillings to over 5,000 feet (1,524 m) got through the reef and terminated in lava. Almost the entire boring penetrated reef and shallow lagoon sediments. Apparently, an old volcano had sunk, and the reef had grown up apace.

Coral banks differ from the other types only in not having a shoal rim. Here, growth seems to have been as good inside the margin as on the rim. Also, inside many atolls, patch reefs have grown to or near the surface. Eniwetok (Fig. 9-5) has numerous patch reefs showing how well the reef may grow inside the atoll rim.

Why Coral Islands?

As we have seen, corals will not grow above the level of low tide. We find, however, that there are hundreds of

Fig. 9-3. Types of coral reef. (See also Fig. 9-5 and Pl. 2C.)

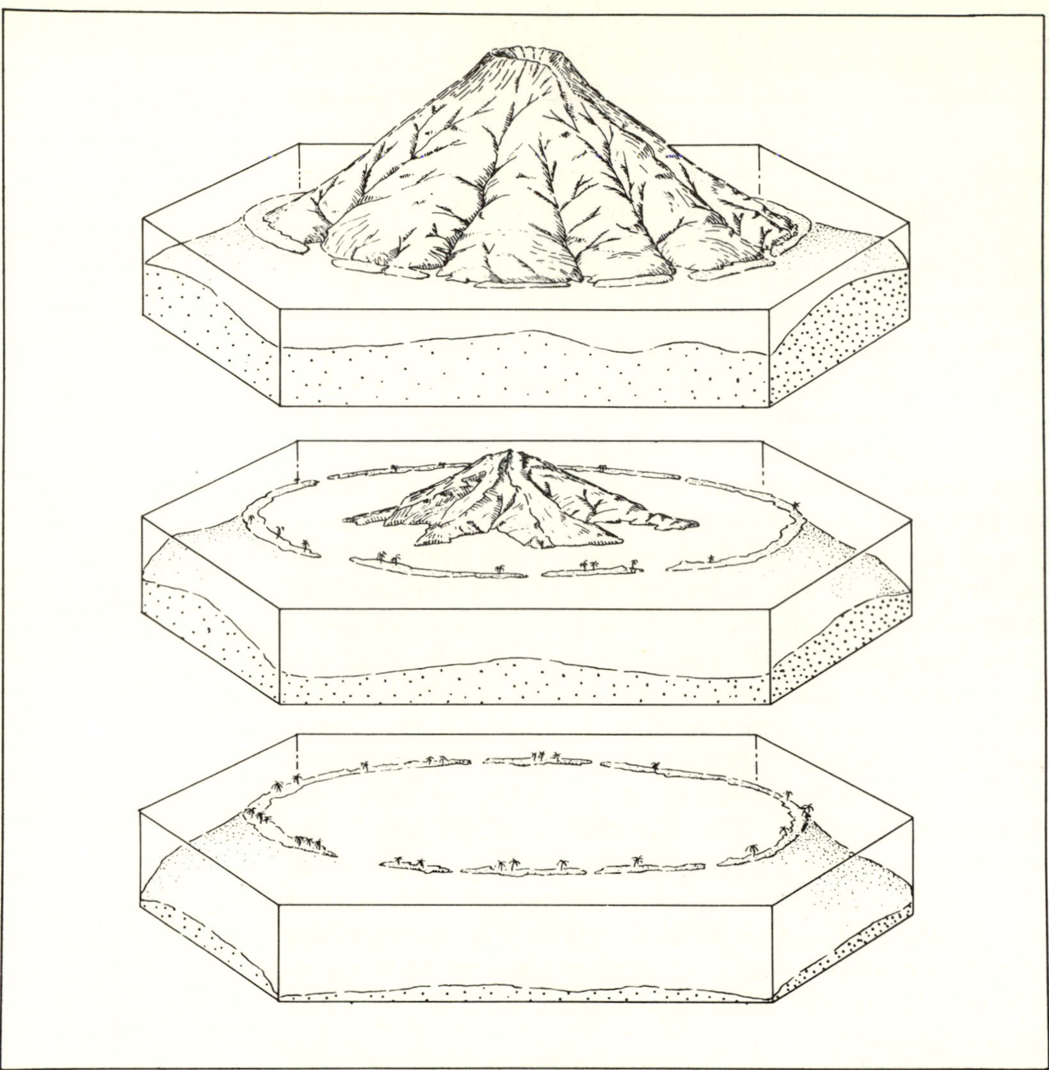

Fig. 9-4. Three stages in the development of an atoll caused by drowning of a volcano and upgrowth of reefs on the margin. Drawn by D. Sayner.

islands that rise above sea level sufficiently to have vegetation, notably palm trees and mangroves. Many islands are inhabited and some have hills rising as much as 100 feet (30 m) above the sea. How is this to be explained?

Once more we are confronted with an explanation that seemed acceptable in the past but is now very much questioned. The old idea was that the present-day sea level is lower than it was a few thousand years ago when the glaciers may have been somewhat smaller. Reginald Daly (1920), author of the turbidity current hypothesis, pointed to numerous terraces around the coasts of the world some 20 feet (6 m) above present sea level. These terraces somehow seemed to shrink in height as studies continued, but are still reported rather commonly up to 10 feet (3 m).

In recent years, very extensive investigations of the late history of coral reefs have been made. The results are not in complete agreement, but in general, scientists have found that the supposed high postglacial sea stands are not well documented. On The *Carmarsel* Expedition of 1967 from Scripps Institution of Oceanography, several

Fig. 9-5. The bottom topography off Eniwetok, Marshall Islands. Darkest shades have depths of 32 to 36 fathoms. Note the numerous patch reefs. From K. O. Emery and others, *U.S. Geological Survey Professional Paper 260,* 1954.

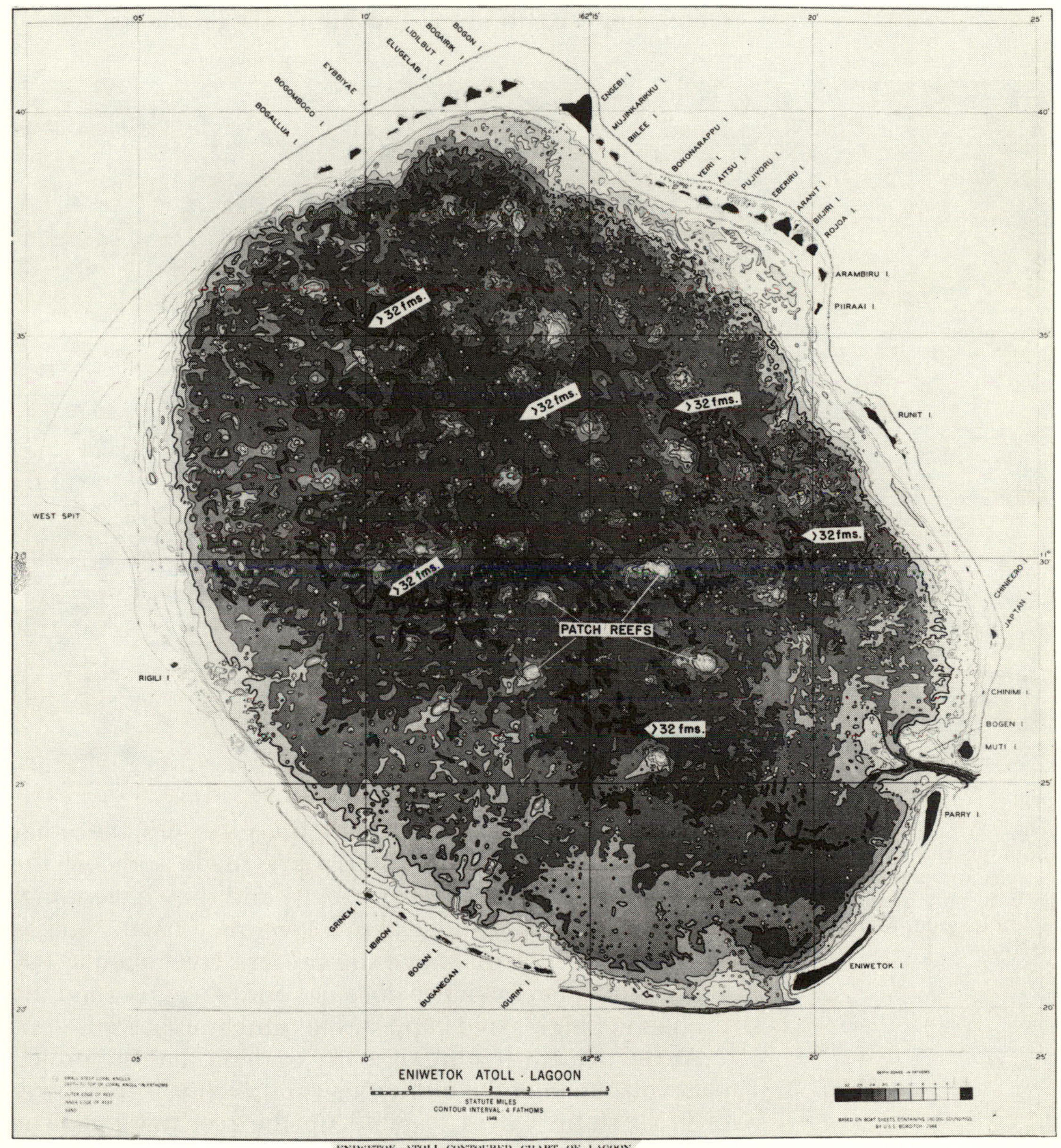

scientists visited 35 islands in what we believe was the most stable portion of the Pacific, which includes the Marshall and Caroline Islands of Micronesia. We found no evidence of the higher sea levels, but instead we found numerous blocks of coral dating as far back as 5,000 years that had been carried up several feet by large storms (Curray et al., 1970) (Fig. 9-6). These blocks make rather even terraces in some places. Other scientists (Thom et al., 1969) have studied the terraces around Australia, which were supposed to show the high postglacial sea levels. They found nothing to support the idea. On the other hand, Tracey and Ladd (1974) have made some of the most careful studies of Pacific reefs, and they have rather good evidence that suggests sea level may have stood as much as 3 feet (1 m) above the present level about 3,000 years ago. Other evidence does not entirely agree, and the problem of high stands appears to remain unsolved.

Fig. 9-6. Masses of coral boulders thrown up by waves onto a reef during two or more hurricanes. Jaluit Lagoon, Marshall Islands.

At the present writing, it is my opinion that the principal explanation of the numerous coral islands is the power of great storm waves to build up the reef margins. The documentation of this idea by Edwin McKee (1958), from a long study of the small island Kapingamarangi, is particularly helpful. He observed the result of a hurricane directly after the storm had ceased. Numerous other dated raised terraces seem to have little consistency in time. They are more likely due to movements of the land than to recent lowering of sea level. The higher hills on coral islands are mostly sand dunes.

Dangers of Swimming on Coral Reefs

The chief reason why tourists fail to get a good look at reefs is probably fear. Undoubtedly some danger does exist. Sharks, as mentioned, are perhaps the greatest deterrent, and barracuda scare many people. Sharks are quite capable of attack but fortunately are seldom aggressive. Barracuda have attacked swimmers but mostly in murky water. If you see a shark, it is best to swim slowly to your boat or to shore, but keep facing the shark because he is probably a timid creature and is much more likely to attack from behind or attack a panicked swimmer who makes great splashes to get away. Chances are you can swim around the inside of a reef for years without seeing a shark. They are more common on the outside.

Other fish are perhaps more dangerous, at least for wounding a swimmer. Some of the large eels are likely to bite someone who puts his hand into a reef crevice where an eel is hiding. Keep your hands out. Other dangers come from stonefish, which look like an oval rock with barnacles on them, and lion-fish, easily recognized by the brilliant colors and long (poisonous) spines. Stepping on either of these will cause terrible suffering and may occasionally cause death. Fortunately, a serum has now been developed which, if available, will cause the pain to disappear in a couple of hours. It is well to avoid cone shells because of their stinging darts. Minor discomfort comes from brushing an arm or leg against fire coral or some types of sea anemone. Again, don't touch things when in doubt. Stepping on the sharp spines of sea urchins can also be very painful. You can watch your step by using a face plate.

Perhaps the most common trouble coming from reef swimming is being pushed by a wave or a current against the sharp edges of many corals (Pl. 3). If you get some coral in a cut it may fester for many days and keep you out of the water.

Swimming inside a reef where you can look at the inner margin is particularly recommended. The waves are usually small. However, look out for a passage leading through the outer reef where there may be strong currents. In fact, watch how you are drifting over the bottom. Currents are usually local and intermittent.

One final suggestion: either wear swim fins or reef shoes to protect your feet from the jagged coral, and wear gloves to protect your hands.

References

Curray, J. R., F. P. Shepard, and H. H. Veeh, 1970. "Late Quaternary sea-level studies in Micronesia: *Carmarsel* Expedition." *Geol. Soc. Amer. Bull.*, v. 81, no. 7, pp. 1865-1880.

Daly, R. A., 1920. "A general sinking of sea-level in recent time." *Proc. Nat. Acad. Sci.*, v. 6, no. 5, pp. 246-250.

Emery, K. O., J. I. Tracey, Jr., and H. S. Ladd, 1954. "Geology of Bikini and nearby atolls." Pt. 1. Geology. *U.S. Geol. Surv. Prof. Paper 260-A*, 265 pp.

Ladd, H. S., J. I. Tracey, Jr., and A. G. Gross, 1970. "Deep drilling on Midway Atoll." *U.S. Geol. Surv. Prof. Paper 680-A*, pp. A1-A22.

McKee, E. D., 1958. "Geology of Kapingamarangi Atoll, Caroline Islands." *Geol. Soc. Amer. Bull.*, v. 69, pp. 241-277.

Thom, B. G., J. R. Hails, and A. R. H. Martin, 1969. "Radiocarbon evidence against higher postglacial sea levels in Eastern Australia." *Mar. Geol.*, v. 7, no. 2, pp. 161-168.

Tracey, J. I., Jr., and H. S. Ladd, 1974. "Quaternary history of Eniwetok and Bikini atolls, Marshall Islands." *Proc. 2nd Int. Coral Reef Symp. 2*, Great Barrier Reef Committee, Brisbane, pp. 537-550.

Suggested Supplementary Reading

Cloud, P.E., Jr., 1964. "Living rock and fossil energy." *The Science Teacher,* v. 31, no. 8.

Hofmeister, J. E., J. I. Jones, J. D. Milliman, D. R. Moore, and H. G. Multer, 1964. *Living and fossil reef types of South Florida. A Guidebook for Field Trip No. 3,* Geol. Soc. Amer. Conv., Nov., 1964, pp. 1-28.

Ladd, H. S., and J. I. Tracey, Jr., 1949. "The problem of coral reefs." *Sci. Monthly,* v. 69, no. 5, pp. 1-8.

Ladd, H. S., and J. I. Tracey, Jr., 1951. "Coral reefs in colour." *The Geographical Magazine,* v. 23, no. 9, pp. 373-383.

Ladd, H. S., J. I. Tracey, Jr., J. W. Wells, and K. O. Emery, 1950. "Organic growth and sedimentation on an atoll." *Jour. Geol.,* v. 58, no. 4, pp. 410-425.

Maxwell, W. G. G., 1968. *Atlas of the Great Barrier Reef.* Elsevier, Amsterdam, 258 pp.

Newell, N. D., 1959. "Question of the coral reefs. Pt. I. Biology of the corals. Pt. II." *Nat. Hist.,* v. 68, no. 3, pp. 119-131; no. 4, pp. 226-235.

Newell, N. D., 1972. "The evolution of reefs." *Sci. Amer.,* v. 226, no. 6, pp. 54-64.

Newell, N. D., J. K. Rigby, A. J. Whiteman, and J. S. Bradley, 1951. "Shoal-water geology and environments, eastern Andros Island, Bahamas." *Amer. Mus. Nat. Hist. Bull.,* v. 97, no. 1, pp. 1-29.

Shepard, F. P., 1976. "Coral reefs of Moorea." *Sea Frontiers,* v. 22, no. 6, pp. 361-366.

Huge canyons that cut the continental slopes

Chapter 10

Geologists have long believed that the large land canyons were cut by swiftly flowing rivers. Because rivers lose their cutting power when they reach the sea, it came as somewhat of a shock when we discovered that the continental slopes are incised with numerous canyons. Even more surprising, we found that submarine canyons include some valleys with the highest and steepest walls of all valleys on the face of the earth (Fig. 10-1). Also, some canyons can be traced seaward to depths of several miles. Before submarine canyons had been extensively explored by marine geologists, the most common explanation for them was that they had been cut by rivers when the lands had locally stood higher than they do at present. However, when their worldwide extent and their enormous depths below adjacent slopes was discovered, many geologists concluded that the canyons had been developed by some marine process, such as turbidity currents, discussed previously. Before giving further attention to the explanations, let's see what the canyons of the sea floor are like. Do they really resemble river canyons?

Thanks to many descents into the shallow heads of the California and French Mediterranean canyons by scuba divers, notably Robert Dill and Maurice Gennesseaux, and in deep-diving vehicles, including dives by Dill, Jacques Cousteau, L. Dangeard, and myself, the general appearance of the canyons is well documented. We have also spent years dredging the canyon walls, taking cores of the canyon-floor sediments, and measuring the currents. Furthermore, a wealth of information has come from the extensive surveys made of the canyons, notably by the U.S. Coast and Geodetic Survey (now NOAA). Off the United States coasts, many submarine canyons are as well surveyed as the canyons of the Rockies and other highlands.

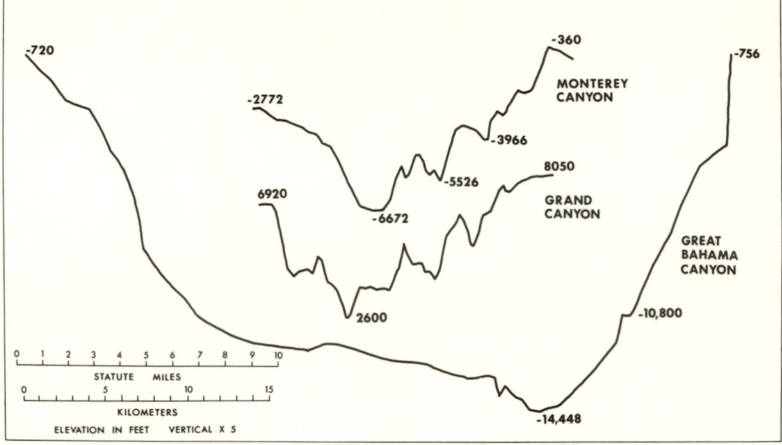

Fig. 10-1. Comparison of profile of the Grand Canyon with profiles of Monterey and Great Bahama Canyons. All profiles use same scale.

Types of Sea-Floor Valley

As a result of the extensive investigations, we have found that there are various types of ocean valley just as there are of land valleys. However, since the best-known valleys of the sea floor have deeply excavated narrow gorges, which have been called *canyons* in land counterparts, the name *submarine canyons* has often been used by scientists for various types of marine valley. To me, this seems unwise. One might as well call the 150-feet deep valley of the Mississippi below New Orleans a canyon as use the same term for the slightly incised valleys cut into the broad fans on the gentle seaward slopes near the deep-ocean floor. Where a series of profiles across a submarine canyon is extended seaward, one usually finds that the inner valley resembles a land canyon, but traced seaward, the walls diminish greatly in height and presently one encounters low bordering ridges, which resemble the levees at the lower end of many river valleys (Fig. 10-2

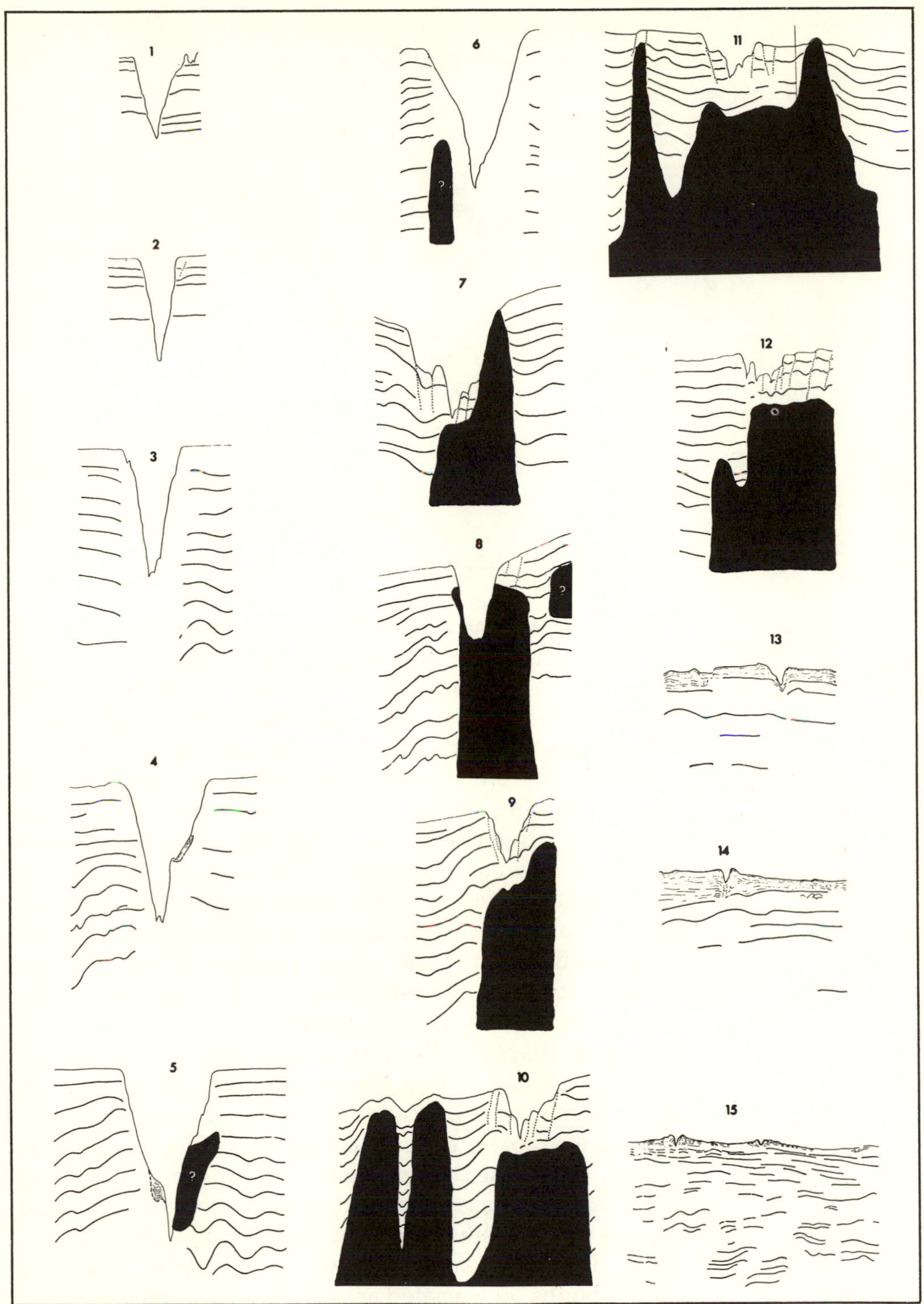

Fig. 10-2. Diagram from seismic profiles across Congo Canyon and fan valley. Solid black represents intruded salt diapirs. Note fan valley profiles 13 to 15.

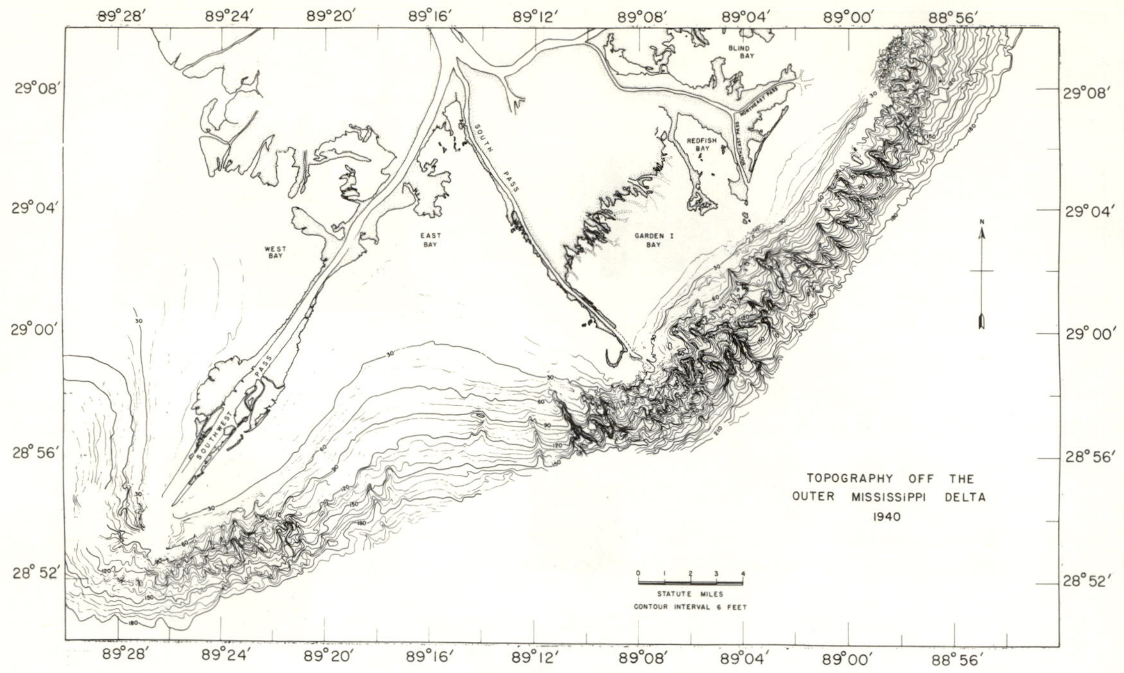

Fig. 10-3. The shallow slope valleys off the passes of the Mississippi Delta. From a survey by the U.S. Coast and Geodetic Survey.

[13-15]). The name canyon should apply to the inner portion, as some of us have suggested; *fan valley* is more appropriate for the lower reaches where the valleys extend across a large submarine fan, somewhat comparable to the alluvial fans at the base of mountain ranges.

Among the other types of marine valley are those cutting submarine slopes where rivers have built deltas across the continental shelf (Fig. 10-3). Although these valleys somewhat resemble canyons, they usually have low walls and they generally differ from typical river canyons in having few tributaries and by extending only part way down a slope. They often have small hills on their outer portions, apparently the result of landslides that seemed to have caused the gullies.

Trough-shaped valleys are found crossing the shelves off a few deltas. These also usually lack tributaries and differ from submarine canyons in having broad flat floors. A good example is off the Ganges Delta (Fig. 10-4). Other trough-shaped sea-floor valleys with rather straight walls are located in areas where there has been extensive faulting, as off the south side of the Aleutian Islands (Fig. 10-5). These appear to be true fault valleys, unlike the shelf troughs off glaciated areas. They resemble Death Valley and the Valley of the Jordan in Israel.

Characteristics of Submarine Canyons

Restricting our use of the word *canyon* to truly canyonlike valleys of the sea floor, we can now explore some of the amazing features that have been uncovered in recent studies. Back in 1934, I was taking wire soundings from a skiff across Scripps Canyon at La Jolla, California. At first, the sounding lead showed some 200 feet (61 m), but then, by moving to the side a mere 10 feet (3 m), the bottom seemed to drop out. Down went the lead, hitting at 450 feet (137 m). We had discovered a vertical wall, and the same thing happened when we sounded down the other side of the canyon. Twenty years later, I was making a descent in Jacques Cousteau's *Diving Saucer* in the same canyon. We followed down one vertical wall, watching it

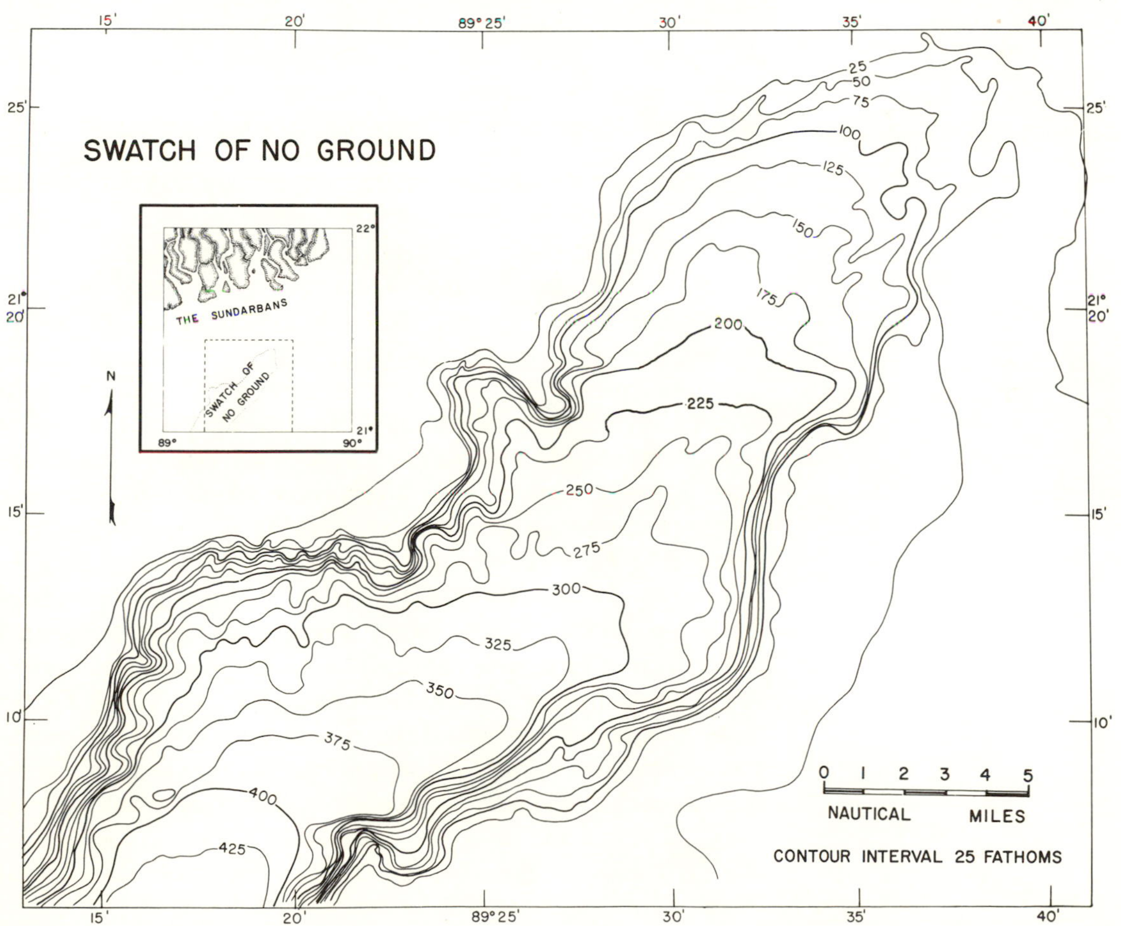

Fig. 10-4. Broad trough valley off the Ganges Delta.

through a porthole. Suddenly the other side of the *Diving Saucer* bumped the other wall. We could not proceed farther; but, tipping the submarine on its side we could just make out the canyon bottom. Elsewhere, cruising along the bottom in the *Diving Saucer,* it was possible to see that the walls were actually overhanging in some places (Fig. 10-6). Inspection of the precipitous canyon walls showed a smoothing and striating of the rock for a short distance above the canyon floor, as if cut by glaciers (Fig. 10-7). Some process of the sea floor was duplicating the effect of glaciers where they move over the lands. However, rising above this smoothed and scratched portion of the wall, we found the rock almost completely covered with marine organisms—small red crabs, deep-living corals, sea anemones, sea urchins, and small mollusks—a veritable rock garden.

Another feature of submarine canyons was discovered by Conrad Limbaugh while diving along a canyon wall at the tip end of Baja California. As you can see from Figure 10-8, it looks as if there was a waterfall at 125 feet (38 m) below sea level. Actually, this is a sandfall and apparently

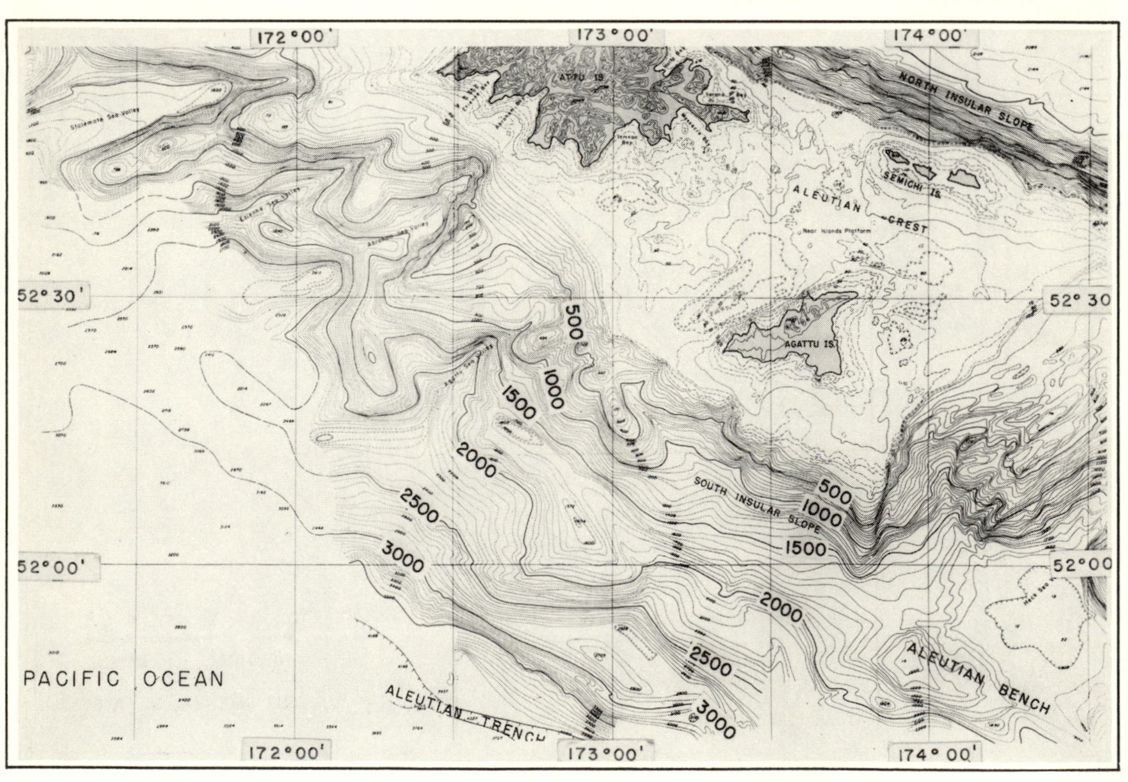

Fig. 10-5. Trough-shaped valleys south of the Aleutian Islands in an area of active earthquakes. Contour interval 50 fathoms. Note the basin depressions. From O. Gates and W. Gibson, *Geological Society of America Bulletin,* 956.

Fig. 10-6. Vertical and partly overhanging rock wall at depth of 880 feet (268 m) in Scripps Canyon, La Jolla, California. Photographed from the *Diving Saucer*.

takes place within canyons when the sand has built up the slopes to beyond the angle of repose so that a slide can occur. Just before this was discovered, a scientific party from Scripps Institution had been camping on a sand fill (a *tombolo*) between two high rock pinnacles (Fig. 10-9), and one night large waves swept across the fill, carrying sand over from the open ocean onto the sloping wall of the canyon inside the point. Later, we were able to descend in the *Diving Saucer* into the same submarine canyon and found on the floor many large freshly broken granite rocks (Fig. 10-10), which had evidently been carried down by these sandfalls landing on a ripple-marked sand-covered floor (Shepard and Dill, 1966; Ch. 5). The ripples indicate still another process that may cause canyon erosion.

In our work at Scripps Institution, we have been measuring the nature of currents in submarine canyons by dropping current meters to the canyon floors and recovering them after a predetermined time interval of days or weeks. An explosive release drops a weight at a known time and allows the attached floats to bring the meters back to the surface (Shepard et al., 1974). The currents, shown on tapes in these meters, usually have been relatively weak, that is, less than one mile (1.6 km) an hour, but on several occasions much stronger currents have carried away the current meters. These losses were always during major storm periods. Two current meters were subsequently recovered during a deep dive. They had been swept downcanyon, and the canyon floor showed clear signs of recent excavation by the currents.

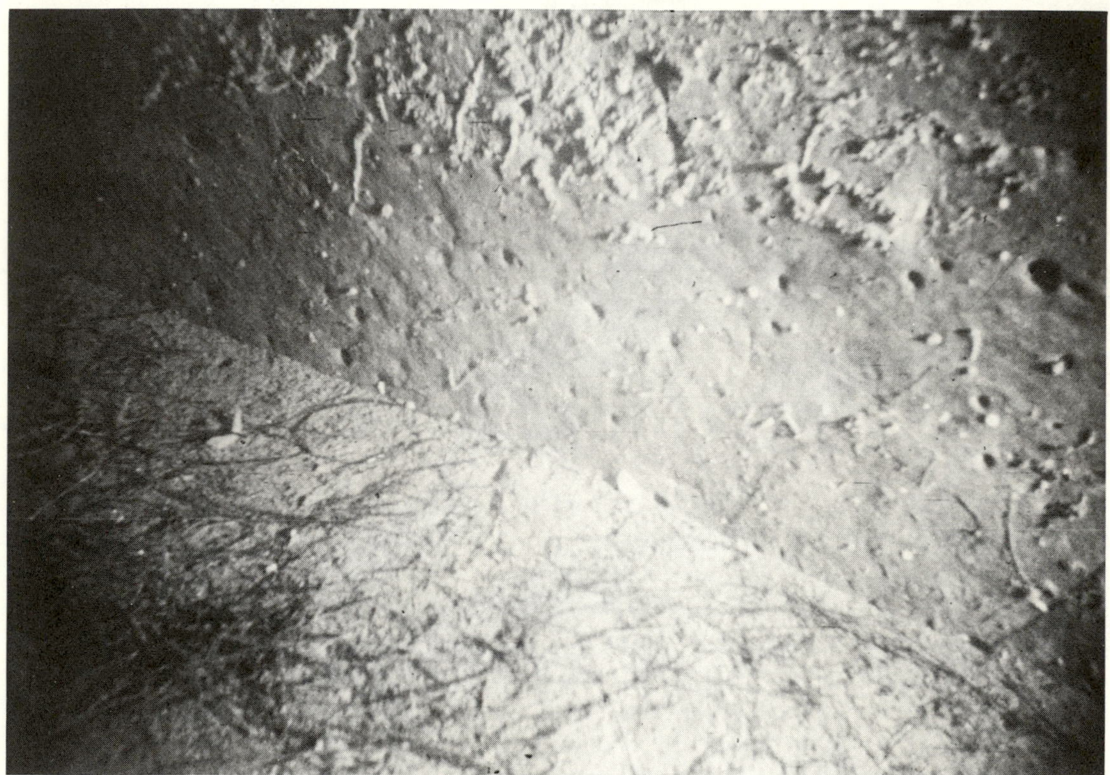

Fig. 10-7. Vertical walls of Scripps Canyon, La Jolla, California, smoothed and scratched. Indentations were formed by pholads and later partly eliminated by erosion of turbidity currents. Canyon floor (foreground) with muddy sand and strands of sea grass. Photo by D. L. Inman from the *Diving Saucer*.

During the dives made by scientists along the walls of submarine canyons, many indications of weathering were observed. Where the walls consist of clays or soft shales, various animals have attacked them, digging holes that resemble the burrowings of animals into land cliffs. The marine burrowers include fish, as can be seen in Figure 10-11 (where the head of a fish is protruding from its burrow). Elsewhere, crabs and lobsters have left the walls looking like a cave-dweller city. Various other animals bore into the rocks. The result is that masses of clay and shale fall frequently down the walls. Even hard rock is penetrated, as in the oval pholad holes shown in Figure 10-7.

The precipitous walls and ripple-marked floors described so far came from depths of less than 300 fathoms (550 m). Do they also occur in the deeper portions of the canyons? The answer is, yes. Photographs taken at depths of more than 2 miles (3.2 km) in a canyon of the Bahama Island area showed both ripple marks (Fig. 10-12A), round cobbles, and apparently steep walls (Fig. 10-12B). In the

Fig. 10-8. Sandfalls on the slopes of San Lucas Canyon, Baja California, following a period when waves swept over the sand beach, shown in Fig. 10-9. Depth 125 feet (38 m). Photo by C. Limbaugh.

Fig. 10-9. Aerial photo showing the beach between large rocks at the end of Cape San Lucas, Baja California. The head of a tributary canyon is outlined by dashed lines. See also Fig. 10-8.

outer part of the Baja California canyon where Limbaugh discovered the sandfalls, we made a dive in the Westinghouse Company's *Deep Star* to a depth of more than 4000 feet (1220 m) and saw both ripple-marked floors and vertical walls of granite.

How deep do the canyons extend? The outer valley continuation of the canyons is a slightly incised fan valley. However, narrow steep-walled true canyons are found seaward to depths of over 10,000 feet (3050 m).

How high are submarine canyon walls? The Grand Canyon of the Colorado has a north wall about 5500 feet (1676 m) high. Both land and sea walls do not have one straight drop but have a series of rock terraces and pinnacles along the way. Great Bahama Submarine Canyon (Fig. 10-1) is 25 miles (40 km) wide where it extends through a passage between Eleuthera and Great Wass Island, and the depth of the canyon floor is 14,448 feet (4400 m). This is more than twice the depth of the Grand Canyon and twice its width. In fact, Great Bahama Canyon may be the deepest canyon in the world.

How long are submarine canyons? Most of them are only a few miles long, but Bering Canyon (Fig. 10-13), in Alaska north of the Bering islands, has a length of 275 miles (442 km), the longest so far discovered. If one includes the outer fan valley, we find that Congo Canyon can be traced for about 585 miles (940 km) but, as seen in Figure 10-3, much of it is not actually a canyon but a fan valley. Some land canyons extend for as much as 1000 miles (1609 km).

Relationship Between Submarine Canyons and Land Valleys

One of the reasons why some of us first thought that submarine canyons were cut by rivers and later submerged was that we noticed that many of the sea canyons headed close to the mouths of land canyons. Congo Ca-

Fig. 10-10. Photo from the *Diving Saucer* of the floor of San Lucas Canyon, Baja California, at 1700 feet (518 m), showing ripple marks with block of granite brought down by sandfalls, like those shown in Fig. 10-8.

Fig. 10-11. Head of a tuskfish protruding from its burrow dug into the wall of Veatch Canyon off New England coast. Photo by R. Slater, courtesy of National Marine Fisheries Center, Woods Hole, Massachusetts.

Fig. 10-12A. Ripple marks on the floor of Great Bahama Canyon at depth of 12,000 feet (3, 658 m).

Fig. 10-12B. Steep rock wall and rounded cobbles near the floor of Great Bahama Canyon, also near 12,000 feet. Compass shows direction. Photo by R. J. Hurley and F. P. Shepard.

nyon actually extends deeply into the estuary of the land valley. Along the west coast of Corsica, a group of submarine canyons enter the embayments and connect with the land canyons (Fig. 10-14). An analysis of all known canyons of the world showed that where the canyon heads were near the coast, 81 percent had a relationship to land valleys (Shepard and Dill, 1966, pg. 231). The canyons lie off many of the largest rivers. This is also true of delta-front troughs.

Some of the canyons that do not show any relationship to land valleys appear to have had a former connection. For example, the canyon off Cap Breton,, southwestern France, now has no river at its head, but in the 15th Century the Adour River entered the sea at this point. Monterey Canyon (Fig. 10-15), the largest off California, has no land valley at Moss Landing, the adjacent shore, but, according to Olaf Jenkins (1973), the slough at Moss Landing may have been one of the outlets of the Great Valley of California during the Pleistocene. An ancient canyon existed here in the Jurassic and has been filled with thou-

A

B

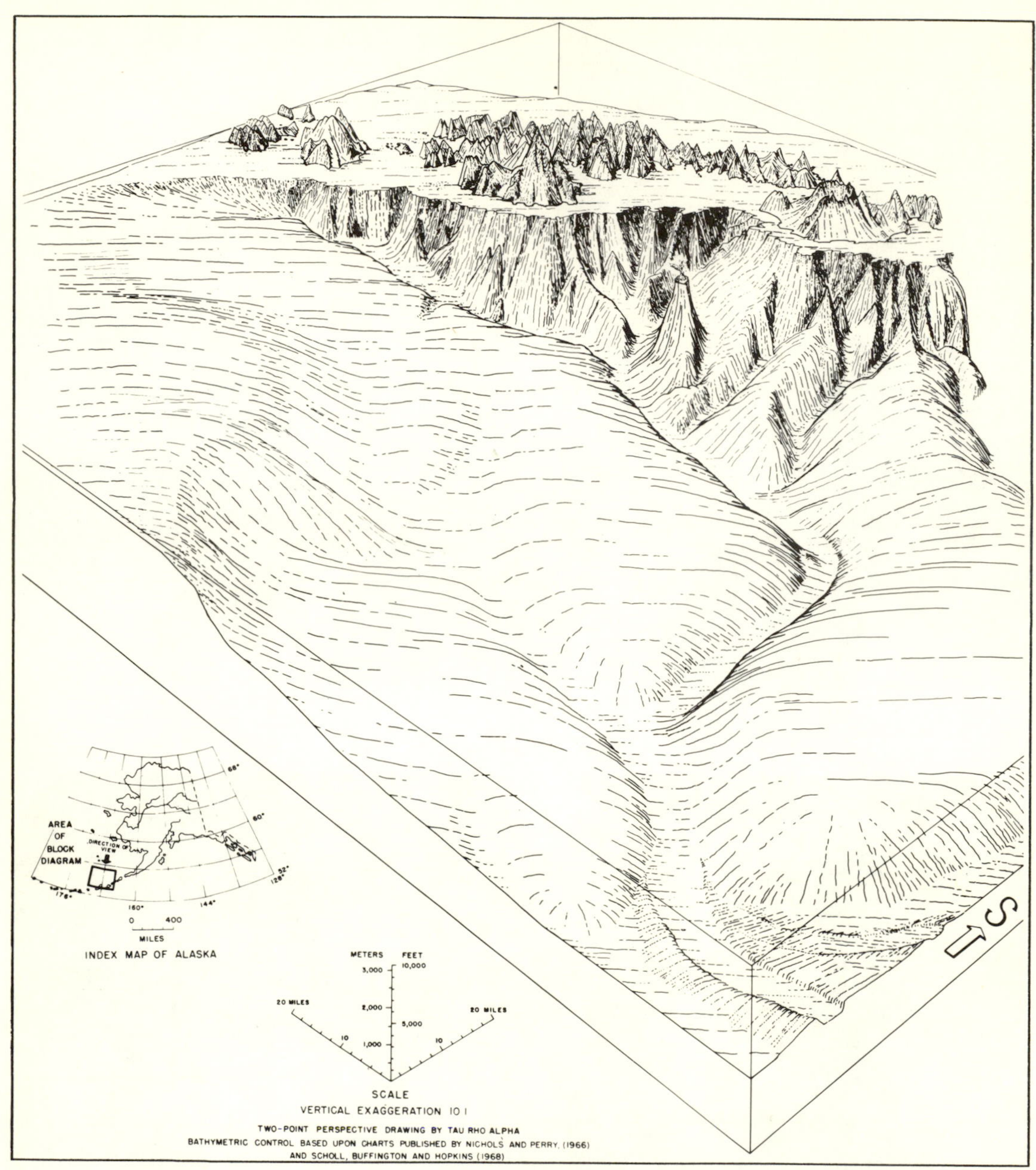

Fig. 10-13. Block diagram of the Bering Canyon in Bering Sea. This is the longest submarine canyon in the world. Diagram by T. R. Alpha, U.S. Geological Survey.

sands of feet of sediment (according to Starke and Howard, 1968).

A few canyons have no known connection with present or old river valleys. One canyon heads in Tanner Bank in the continental Borderland off Southern California. Santa Cruz Canyon, California, heads in a saddle between two of the Santa Barbara Islands, and nearby Hueneme Canyon heads off a projecting point of the Santa Clara Delta (Fig. 10-16), with no sign of an entering river.

Most submarine canyons, even those with the best connections with land valleys, show little resemblance to their land counterparts. The gradient of a submarine canyon is almost invariably much steeper than that of the adjacent land valley. Also, most of the submarine canyons have steeper and higher walls than the nearby land valleys. Other characteristics of submarine canyons include:

- Most of them have winding courses, partly true meanders
- The canyon floors deepen quite consistently seaward
- The canyons lose their deep V-shaped profiles at

Fig. 10-14. Submarine canyons along the west coast of Corsica, which extend into the embayments and connect with river canyons. Contour interval 50 meters. From J. Bourcart and French surveys.

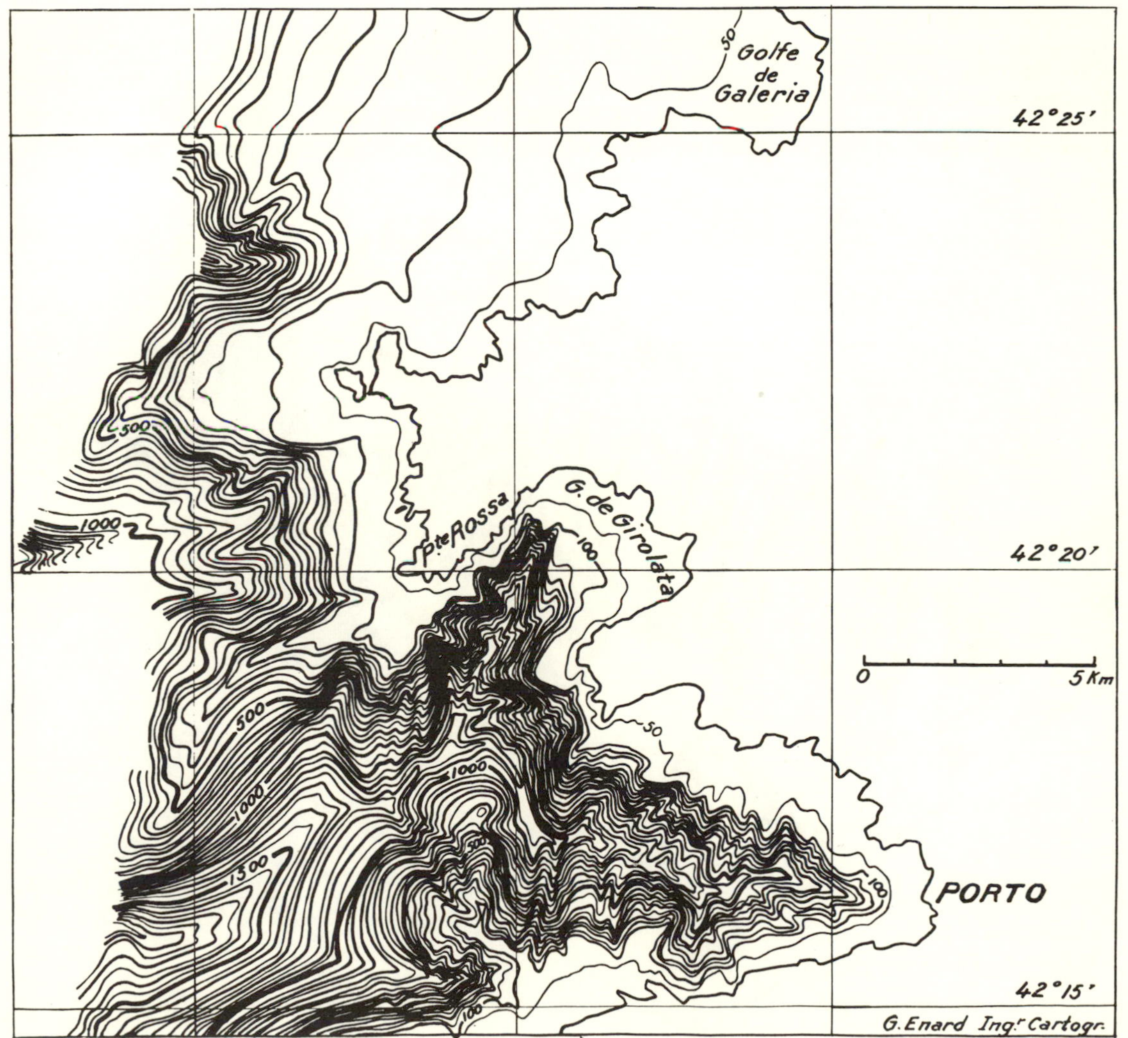

about the same distance from shore as the adjacent continental slope shows a decided decrease in declivity, becoming a part of the continental rise (Fig. 8-1)

- The canyons are rarely found where the continental slopes are gentle
- Canyons are cut into rocks of all degrees of hardness as well as cut into soft sediments
- Most canyons have tributaries, better developed at the canyon heads than in their lower reaches

Origin of Submarine Canyons

Before much was known about submarine canyons, a surprisingly large number of famous geologists tried their hands at explaining the canyons' origin. They were explained as fault valleys, as drowned river canyons, as

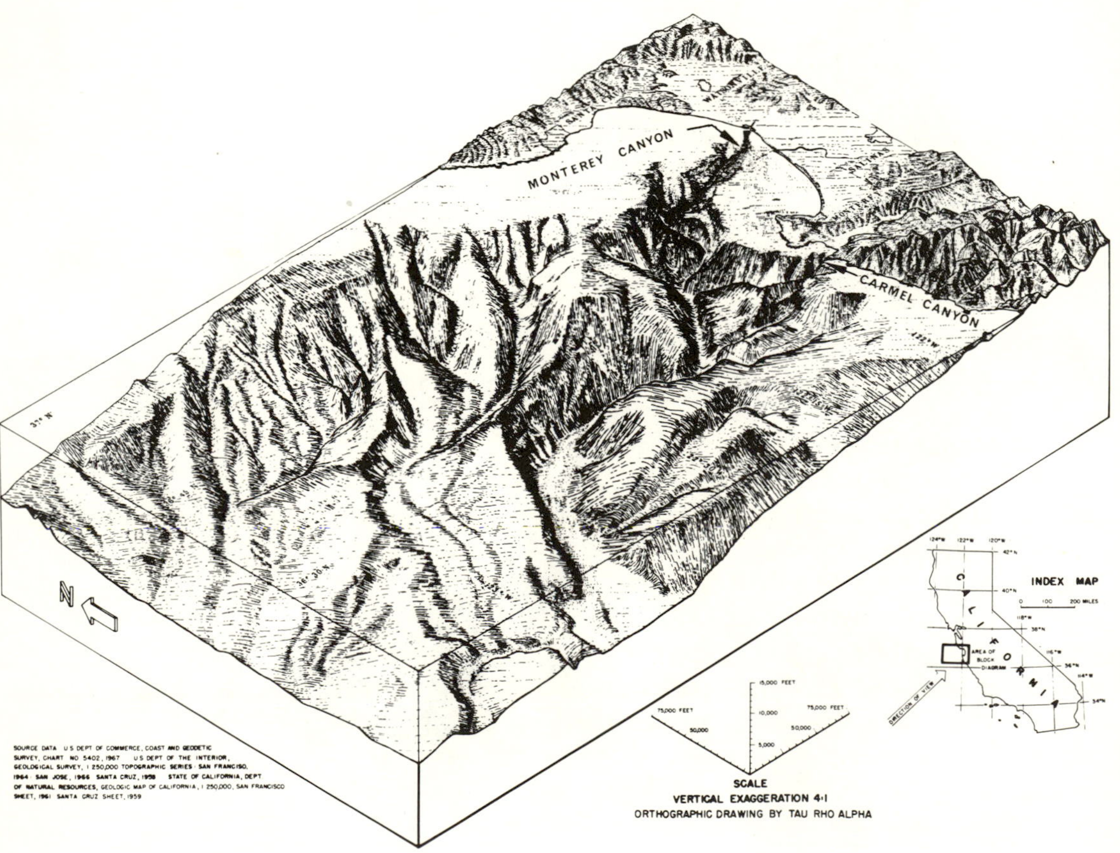

Fig. 10-15. Block diagram of Monterey and Carmel Canyons. Note riverlike pattern and the failure of Monterey Canyon to connect with the river valley; whereas Carmel Canyon connects with Carmel Valley. Diagram by T. R. Alpha, U.S. Geological Survey.

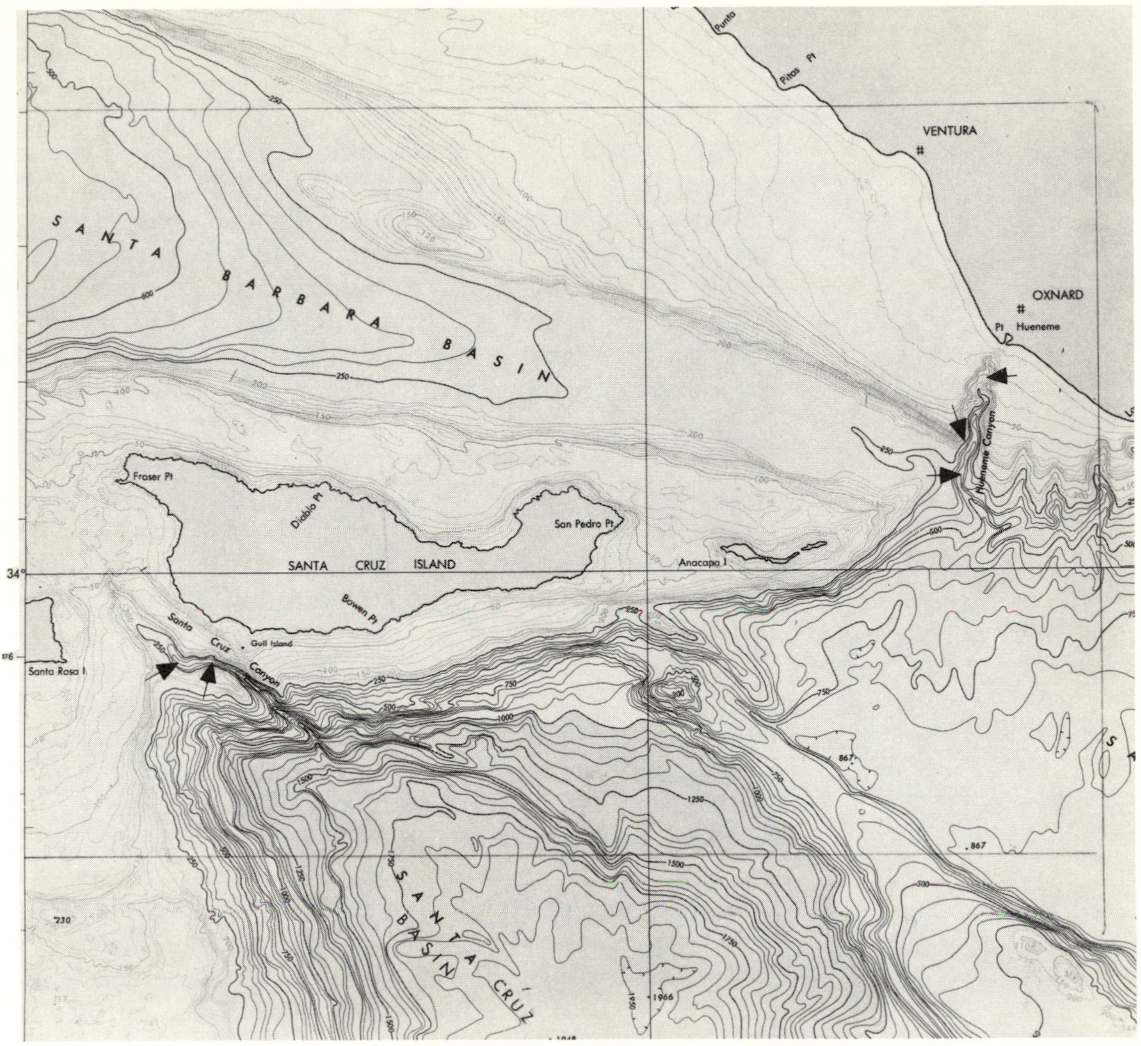

Fig. 10-16. The locations of Santa Cruz Canyon, which heads in a pass between two islands, and Hueneme Canyon, which heads off the point of the Santa Clara Delta, in Southern California. Contour interval 50 meters. Arrows show stations where currents were measured near canyon floors.

due to tsunamis, as the result of the collapse of underground caves, as due to upwelling of deep ocean currents, and as due to turbidity currents. The discovery that canyons are found almost all along the continental slopes caused many people to discard the idea that they were simply due to drowning of land canyons, and the same information proved a very serious weakness in most of the other hypotheses. Thus, tsunamis should concentrate the canyons in a few areas rather than scattering them worldwide. Underground caves should only develop in limestone areas, so why are some canyons cut in granite and others in insoluble rocks? Upwelling is also a local phenomenon and upwelling currents are very weak. Faulting is much more prevalent in some areas than others and faults mostly extend along coasts, rather than directly down the slopes as do typical submarine canyons.

After the Grand Banks earthquake had given much support to Reginald Daly's (1936) idea that turbidity currents were the cause, there was a general swing among geologists toward that explanation. Numerous articles by Philip Kuenen suggesting turbidity currents as the cause also had a profound influence. Our studies at Scripps Institution showed that there were sudden deepenings of the canyon floors (Shepard, 1951). At first we thought these were due to slides at the times of earthquakes, but this was not substantiated. Robert Dill (1964) was inclined to believe that the deepenings were due to slow creep down the canyon floors, and supported his idea by driving in a series of stakes in a line across the floors. The stakes in the center were observed to move faster downcanyon than those of the sides, just like stakes driven into a glacier tongue. However, occasionally all the stakes disappeared. Furthermore, heavy objects, like a car body and large cement blocks emplaced by Dill, were carried away. Finally, with the records of current meters taken during a storm, Douglas Inman and others (1976) observed a current flowing downcanyon at 3 miles (4.8 km) per hour until the wire broke, carrying away the current meter. My group had similar experiences, although we did not get as high velocities indicated in the two current meters we later recovered. During a stormy period, velocities of 2 miles (3.2 km) per hour were recorded by Maurice Gennesseaux and others (1971) in a canyon off the Var River near Nice, France. All these observations strengthened the turbidity-current hypothesis, but unfortunately we still lack good records of any high velocities aside from those of the cable breaks off the Grand Banks.

At the present time, the general tendency is to conclude that submarine canyons are satisfactorily explained as the result of turbidity currents. But this is perhaps premature. There are still some puzzling features that are hard to reconcile with the hypothesis. For example, many submarine canyons are located in areas where the evidence of great submergence during the past few million years seems likely. French geologists have written many articles with supporting evidence for the submergence of the deep canyons along the south coast of France. The borings into the Blake Plateau off the southeastern United States showed that this area has sunk, as has the adjacent Bahama Banks (Goodell and Garman, 1969). Farther north, oil company drillings have indicated that the shelf has been bent down for thousands of feet, and deposits of sediment several miles thick along the outer shelf and slope are shown by sound-wave profiles.

An interesting idea was suggested by Hsu et al. (1973), as the result of their study of the deep borings into the Mediterranean in 1970. They found layers of salt underlying the basins of that sea, and argued that because of its character the salt must have been deposited in shallow water but was later covered by deep-water marine deposits. This sequence seemed best explained by almost complete evaporation of the Mediterranean about 15 million years ago. This might have happened if the Strait of Gibraltar had become blocked and the water supply to the newly formed lake had been less than the rate of evaporation. One of the results of the evaporation would have been that the steep slopes leading to the Mediterranean basin floors would have been laid bare. Hence, streams flowing from the mountainous coasts would have cut canyons into the exposed slopes. This would certainly help explain the great abundance of submarine canyons in the Mediterranean.

Perhaps it is too speculative to consider the possibility that the various canyons along the margins of the Atlantic were also initiated by a similar process, but recent discoveries do show that salt deposits were formed in various places along these margins. An explanation may be that there was a period following the first rifting of the Atlantic during the Jurassic or Cretaceous when the narrow seas in the basins evaporated and exposed to stream erosion the newly formed continental slopes. The objection to the idea is that the slopes must have changed extensively since that remote time so that the canyons could have been filled long ago. However, this argument can be countered by the probability that, once canyons are cut, the turbidity currents keep them open, or at least preserve them as canyons while the shelf margin is being built forward.

One recent discovery about the canyons is that some or perhaps many of them have had two or more stages of cutting, with filling in the intervals. Thus, the sound profile in Fig. 10-17 indicates a canyon cut followed by a fill, and then a later cutting into the fill. In a dive in 1966 in the Woods Hole submarine *Alvin*, Gibson and Schlee (1967) found deposits along the walls of Great Bahama Canyon that they dated as Pliocene, 11 million years old. These were formed in deep water; whereas, oil company drillings into the adjacent Bahama Islands showed shallow-water deposits of the same age at this depth. This they interpreted as indicating that the canyon had existed as a deep-water feature as far back as the Pliocene, again suggesting that submarine canyons may be very old and preserved by turbidity currents.

Once more we seem to have found that a complex origin may apply to submarine canyons. One wonders how many other features have a far more complicated origin than implied by theorists who let their enthusiasms for a simple explanation carry them away.

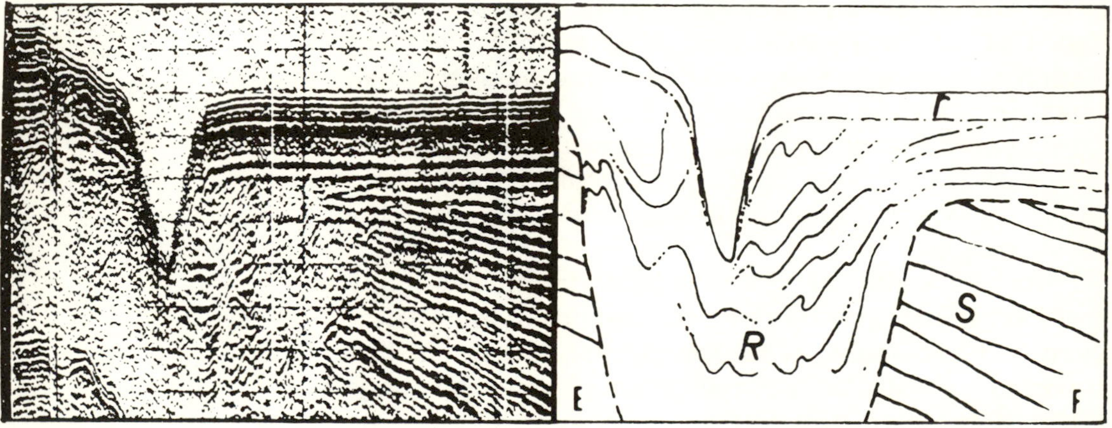

Fig. 10-17. Seismic profile of Cap Crus Canyon in the Mediterranean, southwestern France. Note interpretation on right which indicates two or more episodes of canyon cutting. Courtesy of L. Danegard.

References

Daly, R. A., 1936. "Origin of submarine 'canyons.'" *Amer. Jour. Sci.*, ser. 5, v. 31, no. 186, pp. 401-420.

Dangeard, L., N. Rioult, J.-J. Blanc, and L. Blanc-Vernet, 1968. "Résultats de la plongée en soucoupe no. 421 dans la vallée sous-marine de Planier, au large de Marseille." *Bull. Inst. Oceanog., Found Albert 1, Prince de Monaco,* v. 67, no. 1384, 21 pp.

Dill, R. F., 1964. *Contemporary Submarine Erosion in Scripps Submarine Canyon.* Ph.D. thesis, Scripps Inst. Oceanog., Univ. Calif., San Diego, 269 pp.

Gates, O., and W. Gibson, 1956. "Interpretation of the configuration of the Aleutian Ridge." *Geol. Soc. Amer. Bull.,* v. 67, pp. 127-146.

Gennesseaux, M., P. Guibout, and H. Lacombe, 1971. "Enregistrement de courants de turbidité dans la vallée sous-marine du Var (Alps-Maritimes)." *C.R. Acad. Sci., Paris,* v. 273, pp. 2456-2459.

Gibson, T. G., and J. Schlee, 1967. "Sediments and fossiliferous rocks from the eastern side of the Tongue of the Ocean, Bahamas." *Deep-Sea Res.,* v. 14, no. 6, pp. 691-702.

Goodell, H. G., and R. K. Garman, 1969. "Carbonate geochemistry of Superior Deep Test Well, Andros Island, Bahamas." *Amer. Assoc. Petrol. Geol. Bull.,* v. 53, no. 3, pp. 513-536.

Hsü, K. J., W. B. F. Ryan, and M. B. Cita, 1973. "Late Miocene desiccation of the Mediterranean." *Nature,* v. 242, no. 5395, pp. 240-244.

Inman, D. L., C. E. Nordstrom, and R. E. Flick, 1976. "Currents in submarine canyons: an air-sea-land interaction." *Ann. Rev. Fluid Mechanics,* v. 8, pp. 275-310.

Jenkins, O. P., 1973. "Pleistocene Lake San Benito." *Calif. Geol.,* v. 26, no. 7, pp. 151-163.

Kuenen, P. H., 1953. "Origin and classification of submarine canyons." *Geol. Soc. Amer. Bull.,* v. 64, pp. 1295-1314.

Shepard, F. P., 1951. "Mass movements in submarine canyon heads." *Trans. Amer. Geophys. Union,* v. 32, no. 3, pp. 405-418.

Shepard, F. P., and R. F. Dill, 1966. *Submarine Canyons and Other Sea Valleys.* Rand McNally & Co., Chicago. Ch. 5, p. 231.

Shepard, F. P., and K. O. Emery, 1973. "Congo Submarine Canyon and Fan Valley. "*Amer. Assoc. Petrol. Geol. Bull.,* v. 57, no. 9, pp. 1679-1691.

Shepard, F. P., N. F. Marshall, and P. A. McLoughlin, 1974. "Currents in submarine canyons." *Deep-Sea Res.,* v. 21, no. 9, pp. 691-706.

Suggested Supplementary Reading

Johnson, D. W., 1967. *The Origin of Submarine Canyons.* Hafner Publ. Co., N.Y., 126 pp.

Kuenen, P. H., 1950. *Marine Geology.* John Wiley and Son, Inc., N.Y., pp. 480-525.

Shepard, F. P., 1966. "Submarine canyons and other sea valleys." In *Encyclopedia of Oceanography,* R. W. Fairbridge, ed. Reinhold Publ. Corp., pp. 866-869.

Shepard, F. P., 1974. "Canyons, submarine." In *Encyclopaedia Britannica,* 15th Edit., pp. 786-791.

Shepard, F. P., 1975. "Submarine canyons of the Pacific." *Sea Frontiers,* v. 21, no. 1, pp. 2-13.

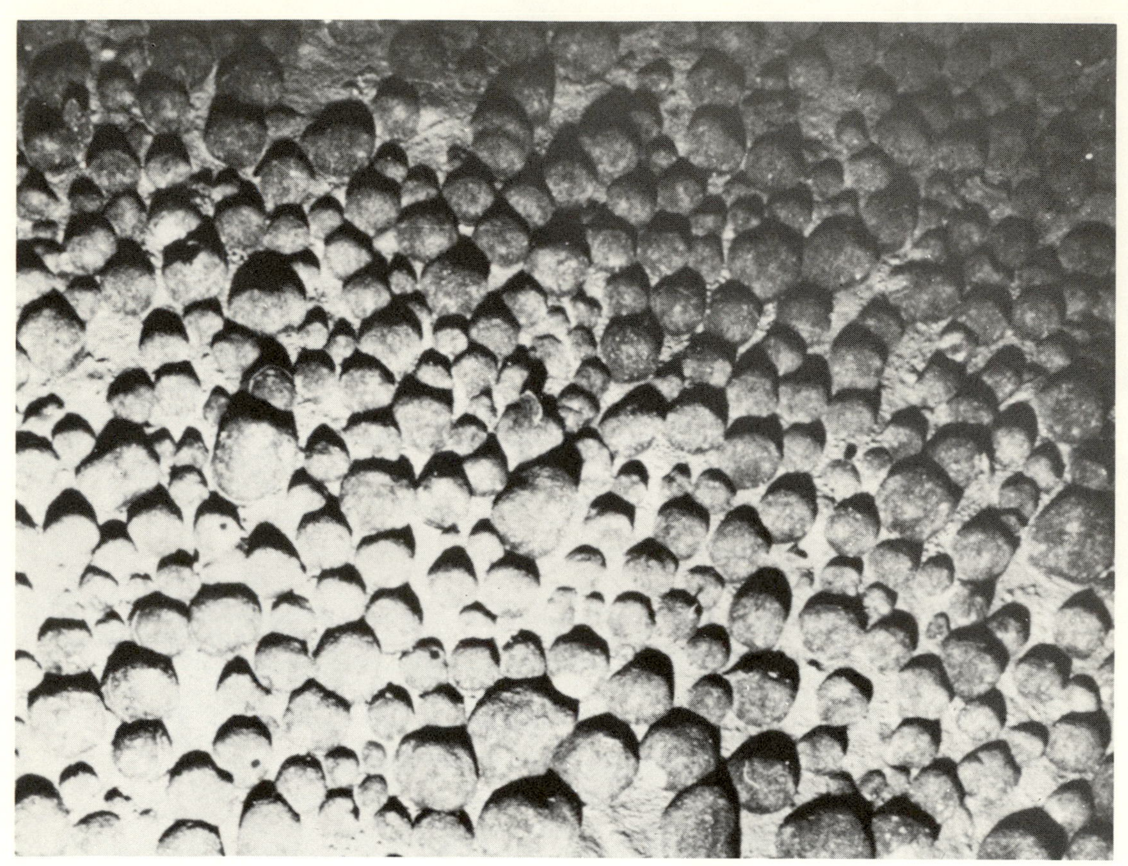

The deep-sea floor

Chapter 11

In this last chapter we will cover a vast territory, about 60 percent of the face of the earth. A few decades ago, this was essentially *terra incognita*, but oceanographers have been concentrating their studies on the deep-sea floor since World War II, and the large Deep-Sea Drilling Project carried out under the direction of Scripps has operated exclusively in deep water. The continental shelf was considered as too likely to encounter oil, causing pollution of the adjacent beaches, like the oil spills off Santa Barbara, California.

For many years after World War II, the navies of the great powers were taking echo-sounding lines across the various oceans and exchanging their results with each other. Up to 1955, attempts to coordinate the lines were discouraging because of poor navigation in mid-ocean and the failure to use constant AC frequency in different machines. Now, with satellite and Loran "C" navigation over much of the ocean and constant frequency in sounding devices of all surveyors, good progress is being made in mapping the deep oceans.

Sound penetration by reflection and refraction has been used widely over the deep-ocean floors, so that we actually know more about the thickness of the sediments in much of the ocean than we do on the continents. At first, the seismic profiles showed only that there were good reflecting surfaces at one or more levels below the bottom, but now with the numerous drillings that have penetrated to these reflecting surfaces we can make good interpretations from many of the profiles.

Topography of the Deep-Sea Floor

In our chapter covering the sea-floor spreading (plate tectonics) hypothesis, we discussed briefly some of the major features of ocean-floor topography. Thus, we mentioned the importance of the Mid-Atlantic Ridge with its rift valley, the great trenches of the Pacific and their relationship to subduction of the ocean crust, and the extensive fracture zones that offset all of the mid-ocean ridges. In the chapter on continental shelves, we defined the continental rise. Now we can examine more closely each of these features and discuss the extensive seamounts that have been discovered rising above the deep-ocean floor.

Continental rise. The gently sloping floor at the base of the steep continental slope, already referred to as the continental rise, is found on both sides of the Atlantic, along portions of the Indian Ocean, and around much of Antarctica but rarely along the Pacific margins (Fig. 11-1). This distribution is by no means haphazard. It is clearly found along what is thought to represent the torn edges of the ancient continent of Pangaea, believed to have existed be-

Fig. 11-1. Global distribution of continental rises. Note that they are almost all where the continents have been pulled apart, according to the sea-floor spreading hypothesis. From K. O. Emery, *Oil and Gas Journal*, 1969.

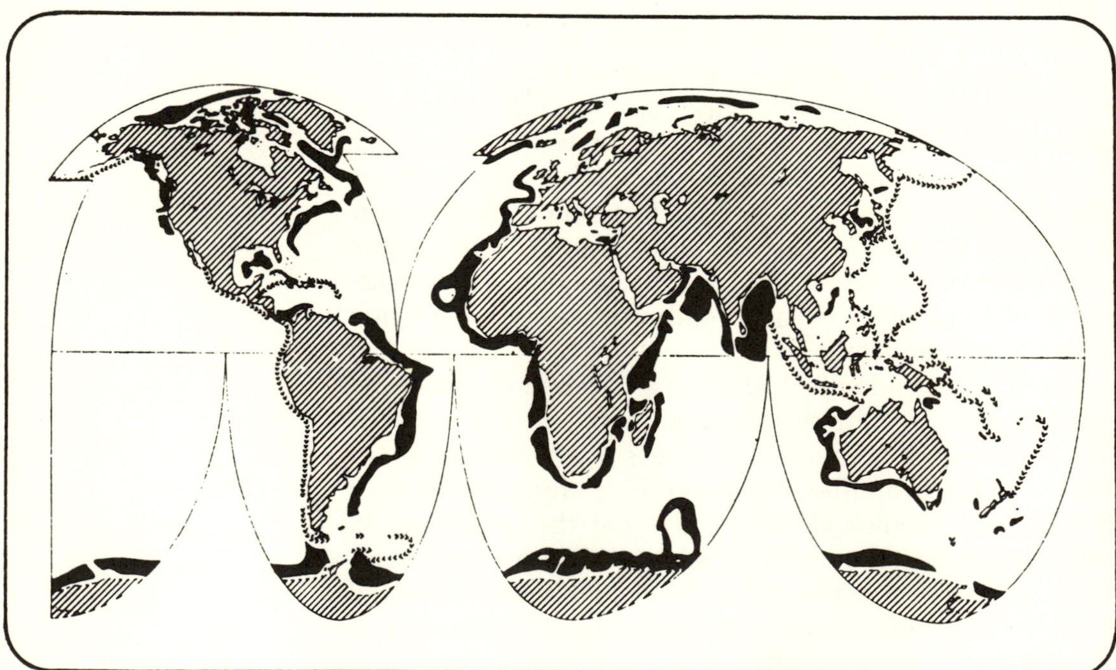

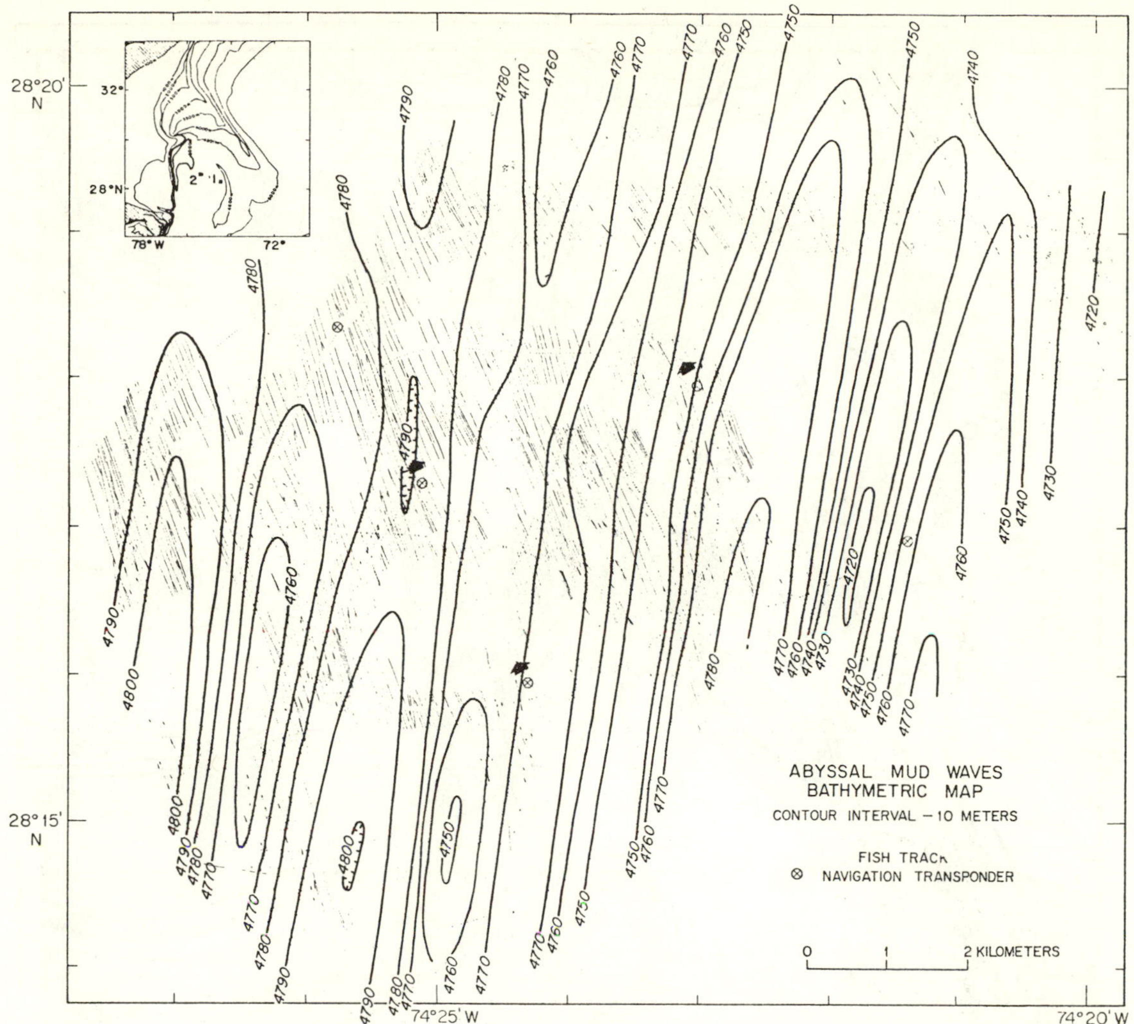

Fig. 11-2. Abyssal mud waves found in deep water north of the Bahama Islands. Interpreted as coming from the north and transported by abyssal currents. From C. Hollister and others, *Geology*, 1974.

fore the general continental breakup in the Jurassic (Fig. 2-1). As the Atlantic, Indian, and Antarctic Oceans grew, the trailing edges must have received sediments built out onto the margins of the new oceans from the adjacent land masses. This seems to be a satisfactory explanation for the great overlapping fans of sediment that constitute the continental rises. Their absence along most Pacific coasts also fits the sea-floor spreading hypothesis because it was these margins where, according to the hypothesis, the oceanic crust was being carried down under the advancing continental plates so that the sediment fans would have been engulfed, leaving steep slopes and ocean trenches next to the continents.

Like many other features of the sea floor, the continen-

Fig. 11-3. Mid-ocean ridges in the Indian Ocean and the extension of one arm into the Gulf of Aden, Red Sea, Jordan Valley, and the rift valleys of Africa. From B. Heezen and M. Tharp, *Geological Society of America Special Paper No. 65.*

tal rises cannot be entirely explained by this one hypothesis. They are not just great sediment fans. Along the southeastern margin of the United States, the continental rise has some hills and ridges rising above the general floor (Fig. 11-2). Seismic profiles show that these represent sedimentary accumulations, like giant sandbars, and are probably deposits from the relatively strong bottom currents that sweep in a southerly direction along the continental rise.

Mid-ocean ridges. The ridge that extends down the Atlantic from north to south virtually follows the center of that ocean, but its continuation around Africa and up into the Indian Ocean has a more complicated pattern (Pl. 1). Two converging arms of the Indian Ridges (Fig. 11-3) extend up

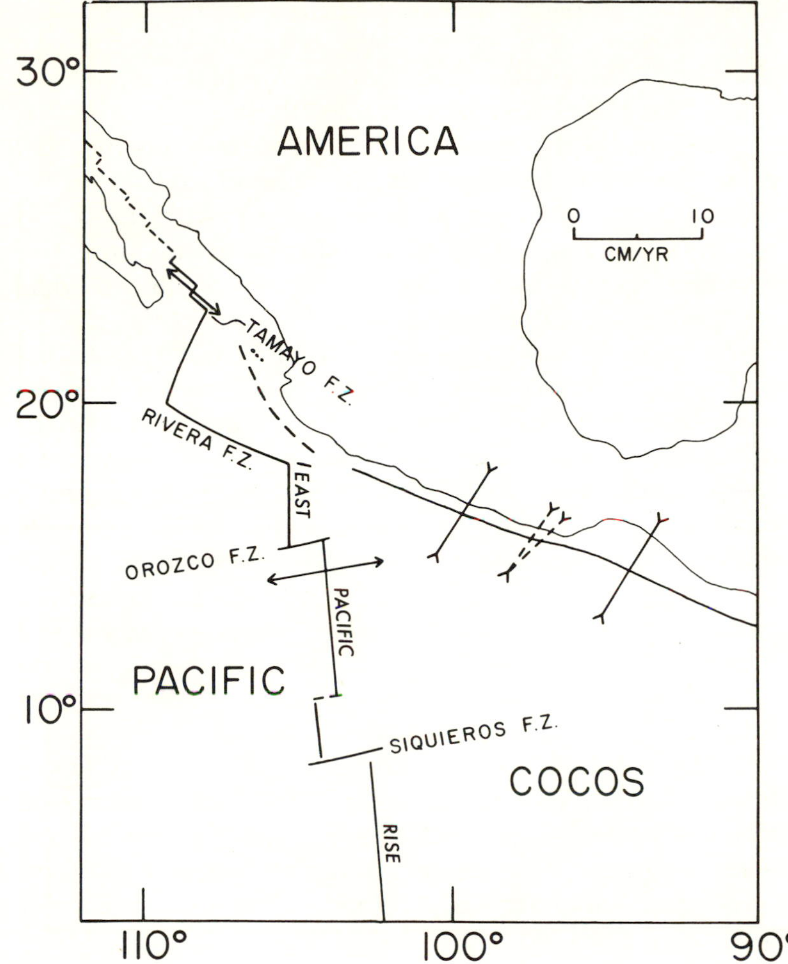

Fig. 11-4. Showing how the East Pacific Rise extends up into the Gulf of California. Rates of spreading are indicated by arrow lengths. Note the offset in the crest of the rise caused by fracture zones. From R. Larson, 1970.

into the Arabian Sea and continue north of the convergence as an inverted Y with a single ridge terminating in the Gulf of Aden, where it appears to become a rift valley bending up into the Red Sea and connected with the Rift Valleys of Africa, to the south, and the Jordan-Dead Sea valley, to the north. The southeastern arm of the inverted Y extends east, passing south of Australia and then crossing the Pacific Ocean as the East Pacific Rise in a northeasterly direction until it comes to the entrance of the

Gulf of California, another rift valley (Fig. 11-4). Thus, we have two examples of the Mid-Ocean Ridge coming into the land and, in both cases, rifting action is clearly indicated. However, along most of the Indian Ridge and of its continuation into the East Pacific Rise there is almost no rift valley along the axis of these ridge continuations. In most other respects the world-encircling ridge is similar. The matching belts of magnetic anomalies on the two sides of the Mid-Atlantic Ridge are even better developed in the East Pacific Rise than in the Atlantic.

The name Mid-Ocean Ridge is not entirely justified because two of these ridges actually extend into the continents (Pl. 1). However, as was stressed by Menard and Smith (1966), most of the ridges show a good symmetry, dividing sectors of the oceans into two almost equal portions.

The fracture zones. It was only after World War II that hydrographers and marine geologists discovered the existence of great fracture zones that cut the mid-ocean ridges into many sectors (Fig. 2-5). The east-west Romanche Deep, off the west coast of Africa, had been known previously but no one thought of this as part of an extensive pattern of ridges and troughs that involves all of the oceans and extends across all the mid-ocean ridges. The pattern finally began to appear as the result of the work of Menard (1964) in the Pacific off the California coast. At first, Menard and Dietz (1952) explored the great east-west ridge off Cape Mendocino, which has an escarpment a mile (1.6 km) high on its south flank. Then Menard (1964) found several other east-west ridges and associated troughs farther south, and he suggesed the term *fracture zone* for them. The explorations by Lamont-Doherty Geological Laboratory and Woods Hole Oceanographic Institution soon led to the discovery of many more fracture zones all over the world. At first, the fracture zones of the northwest Pacific were not associated with the newly developed sea-floor spreading hypothesis, because it was not appreciated that sea-floor spreading was applicable to the Pacific, particularly north of where the East Pacific Rise terminates at the Gulf of California. Soon it was pointed out by Victor Vacquier (1972) of Scripps Institution that the magnetic anomaly belts running at right angles to the fracture zones were greatly offset along them. The same thing was found by Fred Vine (1968) and many others for the fracture zones in the Atlantic and in the Indian Oceans.

As ocean exploration continued, we were all surprised to find that fracture zones are by no means local features cutting the mid-ocean ridges, but in many places can be traced for thousands of miles on each side. It is still hard to understand how these features of great relief could have been missed for such a long time after echo sounding had been introduced (see Fig. 2-9B). However, most ocean soundings were taken by ships crossing the oceans, rather than running parallel to the coasts where the fracture zones would have been encountered.

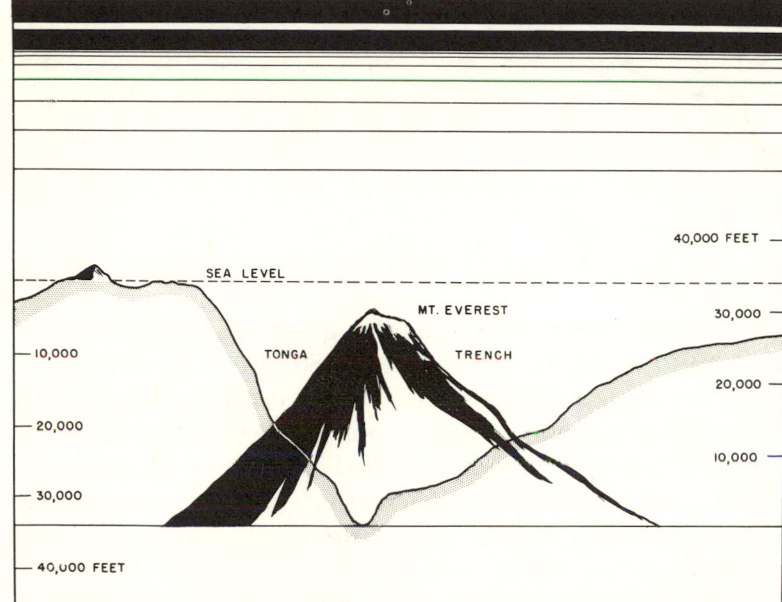

Fig. 11-5. Comparison between the Tonga Trench and Mt. Everest. Sections have the same scales.

Deep-sea trenches. The greatest depths in all of the oceans are found in the trenches, mostly along the continental margins of the Pacific (Fig. 2-4). The three deepest are all in the western Pacific: the Mariana Trench with a depth of 35,700 feet (10,881 m), the Tonga Trench with 35,370 feet (10,780 m), and the Mindanao Trench with 34,420 feet (10,494 m). A comparison between the Tonga Trench and the highest mountain on the continents (Mt. Everest) suggests the major relief of the sea floor is of the same order of magnitude as that on the land (Fig. 11-5).

Although the trenches off the Americas are not nearly as deep as those of the western Pacific, if one adds the height of the Andes to the 28,254 feet (8612 m) depth of the

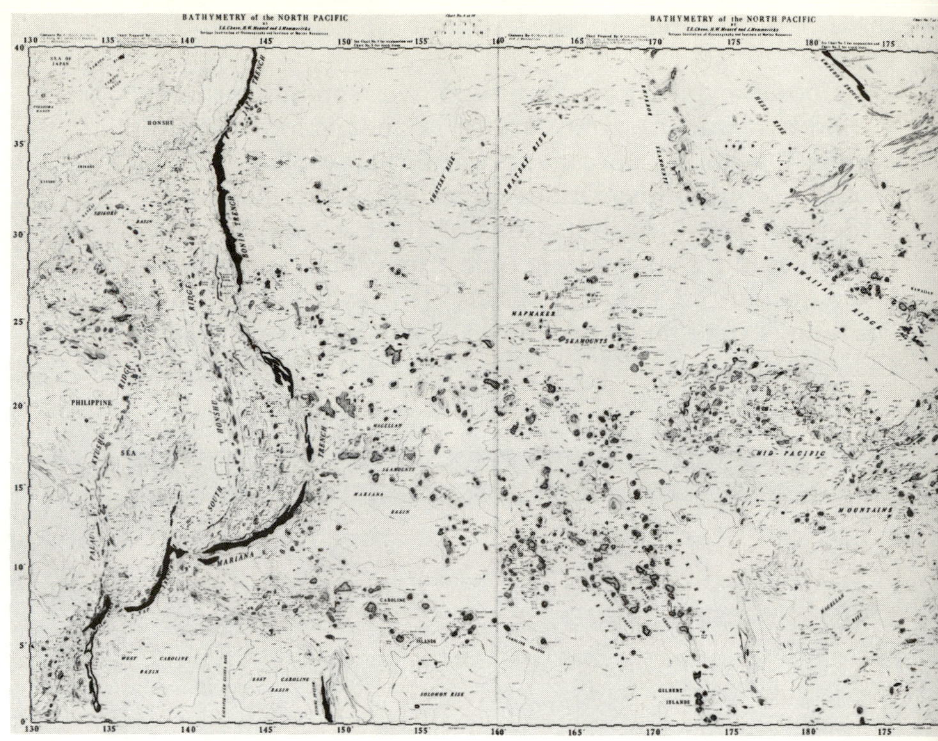

Fig. 11-6. Bathymetry of the North Pacific with the numerous fracture zones, abundant guyots, and seamounts. From T. Chase, H. W. Menard, and J. Mammericks.

Peru-Chile Trench one finds the greatest known relief in the world is about 42,000 feet (12,802 m). The two small trenches in the Atlantic that partly encircle the arcuate West Indies and the arcuate Scotia Ridge (South Sandwich Islands) apparently do not exceed 30,000 feet (9,144 m) in depth, and the Java Trench that curves around the Indonesian arc, has a maximum depth of 24,390 feet (7434 m).

The trenches are partly V-shaped in cross section and partly flat floored. One of the surprising features of the trenches, first pointed out by David Scholl and others (1970), is the discovery that the sediments under the flat floors are generally horizontal. This is contrary to what one might expect if they have been bent down by the subduction of the ocean crust. Explanations for this inconsistency seem somewhat inadequate, but the question can only be solved by more investigations. All geophysicists now agree that there is a large deficiency in gravity under the trenches, called a *negative anomaly*, that favors downbending (Fig. 2-4). Also, earthquakes are common in trenches all around the Pacific and are often followed by

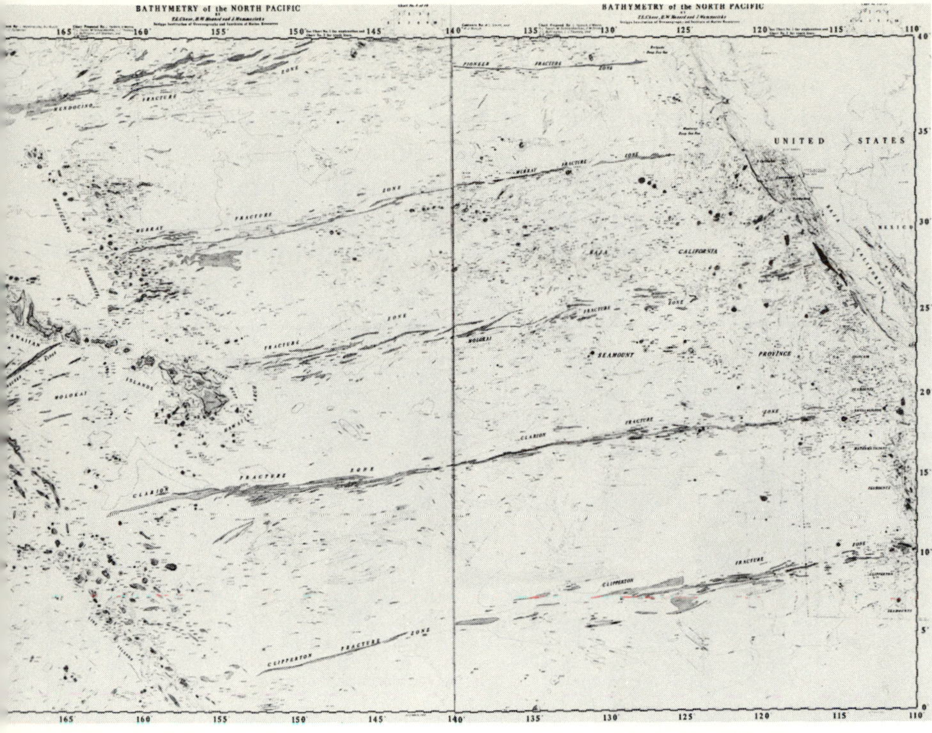

large tsunamis, showing that these deeps are still in the process of formation.

Seamounts and guyots. Seamounts are defined as "isolated sea-floor elevations rising to 3000 feet (914 m) or more above their surroundings." When these have flat tops they are called *guyots*, a name suggested by Harry Hess. During World War II, Hess was a Navy captain who made numerous Pacific Ocean crossings to the war zones, keeping an echo sounder running continuously. He discovered an abundance of previously unknown seamounts and guyots (Hess, 1946). Many new ones have been found since as the Pacific became better explored (Fig. 11-6). Most of these undersea mountains are along lines, many of them in fracture zones, but others are along the volcanic ridges and are usually explained as formed by the ocean crust moving over hot spots and forming rows of volcanoes (see Chapter 2). The other oceans do not have as many of these seamounts as the Pacific, but new ones are frequently discovered.

The usual explanation of the guyots is that they represent drowned wave-beveled platforms or former coral banks. We have some evidence favoring both of these ideas. Edwin Hamilton (1956) found that an underwater mountain range southwest of the Hawaiian chain had drowned coral reefs of Cretaceous age. Matthews and others (1974) found ample evidence that other Pacific guyots are old platforms built to sea level by coral growth prior to submergence. The drillings into coral reefs, mentioned previously, have also shown evidence of the great

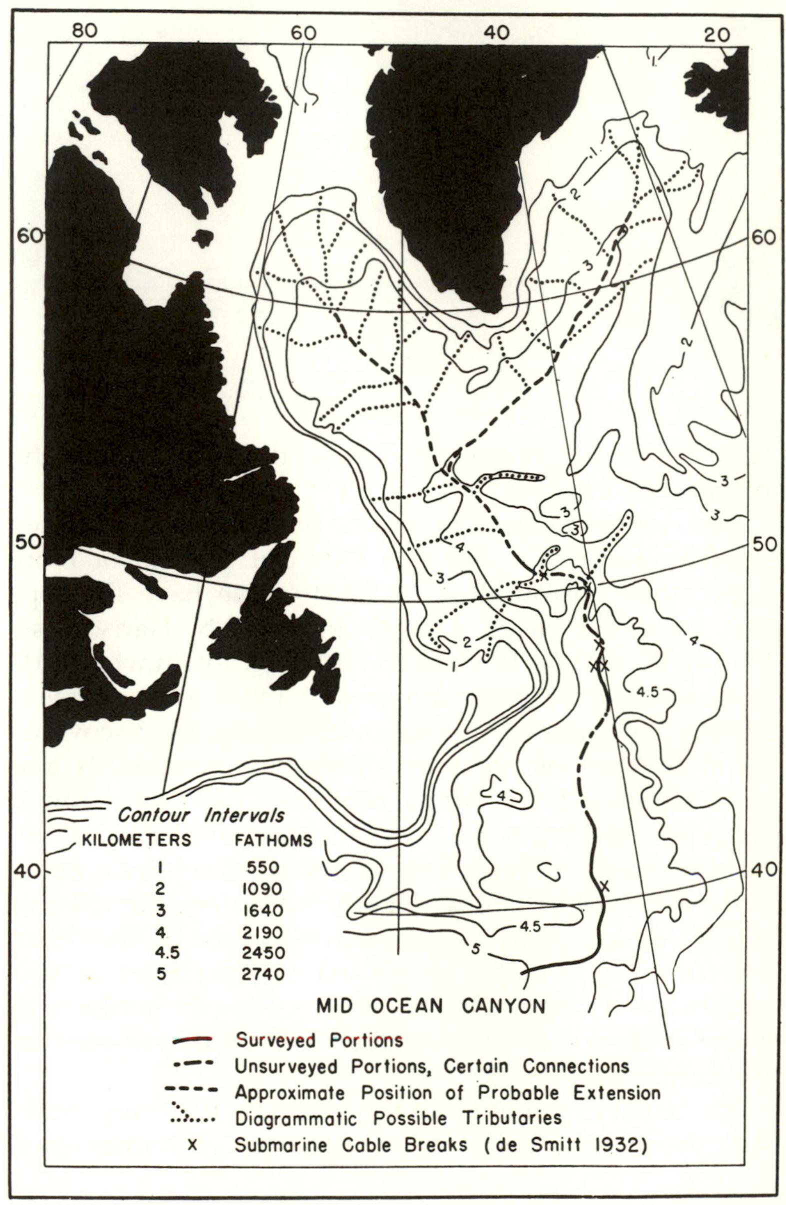

Fig. 11-7. The Mid-Ocean "Canyon." The tributaries are speculative. This trough-shaped deep-sea channel is thought to be due to turbidity currents. From B. Heezen and others, *Geological Society of America, Special Paper 65,* 1959.

submergence of ancient platforms. It would be exciting to learn that some of these submerged platforms had supported ancient civilizations, like the fabled Atlantis and the Continent of Mu. Unfortunately, no exploration has given any support to these tales. The submergence of guyots appears to date back well into the Tertiary or Cretaceous, long before the coming of man.

One should consider another possibility for at least some of the guyots. Their flat tops could be of volcanic origin. Some land volcanoes have flat platforms, the result of lava filling old craters. Submarine craters might also be filled by deep-sea deposits, which could become solidified and more resistant to underwater erosion than the lava breccia or volcanic dust that form some of the crater rims.

The idea suggested by Hess (1946), that the guyots of the Pacific might represent a general sinking of the sea floor and rising of sea level since Pre-Cambrian times (a billion years ago), is now refuted by the dating of the rocks dredged from the guyots, but some submergence of these platforms has occurred in more recent ages as the result of the extra weight on the oceanic crust of the volcanic mountains. The ocean crust becomes deeper as it becomes older due to thermal contraction. Whatever the cause, the guyots seem to have been ideal platforms on which the large Pacific coral atolls could be built as the land sank.

Deep-sea channels. One of the mysteries of the deep ocean comes from the discovery of the slightly incised channels winding down the ocean floors toward the deep central basins. Some, or perhaps all, of them are continuations of the fan valleys at the lower ends of submarine canyons. These channels are mostly trough-shaped with steep sides and fairly flat floors. In general, they deepen consistently toward the ocean basins. They may have levees on the sides, like the fan valleys, although this has not been well confirmed as yet. In some places, these channels cross fracture-zone ridges, becoming locally true canyons. Sampling of the floors of the channels shows that they have sand layers and even gravel deposits.

One of the best known of these channels, explored by Heezen and others (1959) winds out of Baffin Bay, around the Grand Banks, and terminates in the Nares Deep at 2700 fathoms (4938 m) (Fig. 11-7). Farther south along the equator, a partly filled "mid-ocean canyon" appears to have drained westward from the Brazilian continental rise towards the center of the Atlantic Ocean (Damuth and Gorini, 1976) but was later largely filled with sediments. Further exploration will probably reveal that many more features of this sort have now been buried. Cascadia Channel, heading off the Strait of Juan de Fuca (Fig. 11-8), has been explored by Robert Hurley (1960) and others (Griggs et al., 1969). Again the channel changes its course, in this case from southwest to south, so that both of these well-known channels extend mostly parallel to the coast.

The channels of the Bay of Bengal should perhaps be included here, although in the upper bay they are clearly fan valleys and partly connected with the delta-front trough off the Ganges-Brahmaputra River system (Fig. 11-9). The explorations of Curray and Moore (1971) have shown that a network of channels extends down the entire bay and can be traced out into the Indian Ocean to latitude 5°S. With the cooperation of Russell Raitt, they found that the channels overlie the thickest known mass of sediment on the sea floor, 50,000 feet (15,240 m). The gradients are extremely low, about 10 feet per mile (1.7 m per km). Other channels are found on the sea floor off southern Alaska and south of the British Isles.

The explanation for channels of the deep-sea floor is the same as the most prevalent explanation for submarine canyons. Turbidity currents seem to be the best answer at the present writing. If this is correct, these currents show a remarkable ability to continue down extremely gentle slopes. How they can cross the ridges of the fracture zones and what process can even transport gravel along them is a real puzzle, but at present there seems to be no good alternative explanation.

Deep-Ocean Sediments

Geologists have had a fairly good idea of the nature of sediments on the deep-ocean floor ever since the British *Challenger* Expedition of 1872-76. Extensive reports emerged over a long period, well summarized by Murray and Renard (1891). They found in the short cores of the expedition that the deep-ocean floor had two predominant types of deposit. In the deepest water, clay was found

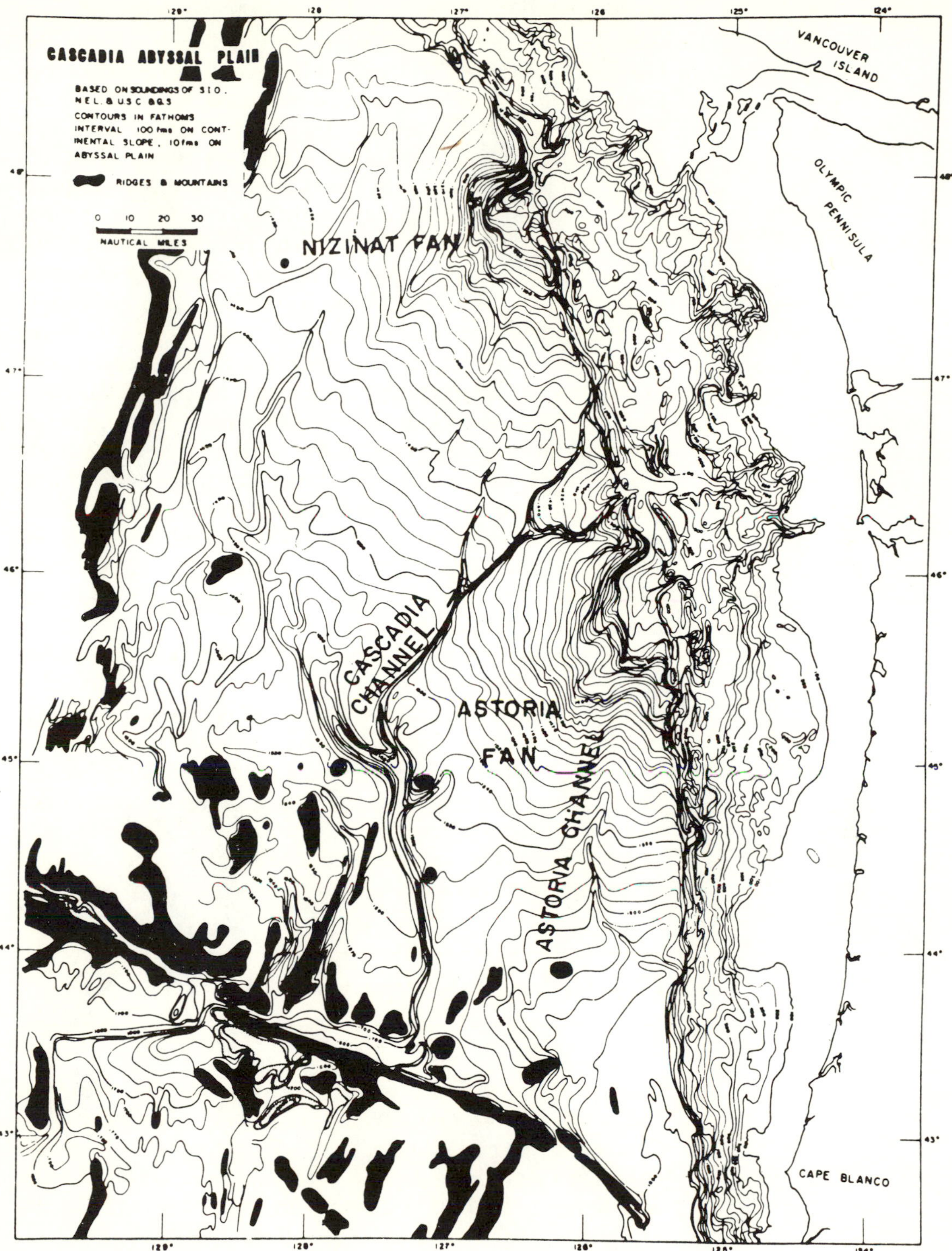

Fig. 11-8. Cascadia and Astoria Channels heading off the Columbia River and Astoria Fan. Note that one channel crosses submarine ridges and extends to great depths of the ocean floor. From R. Hurley, unpublished Ph.D. thesis, University of California, San Diego.

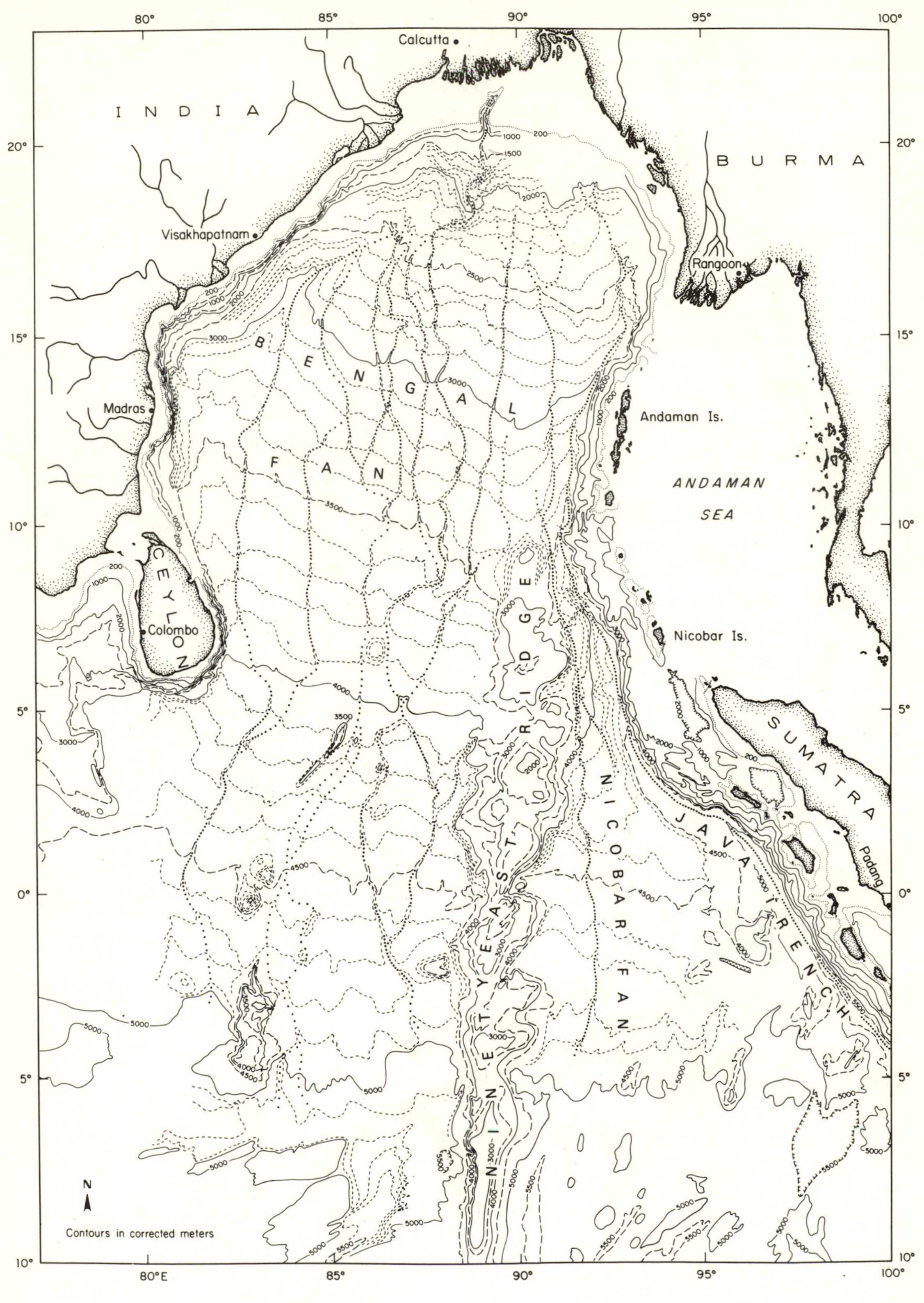
INDIA
BURMA
Calcutta
Visakhapatnam
Madras
Rangoon
Colombo
CEYLON
BENGAL FAN
Andaman Is.
ANDAMAN SEA
Nicobar Is.
SUMATRA
Padang
NINETYEAST RIDGE
NICOBAR FAN
JAVA TRENCH
N
Contours in corrected meters
80°E
85°
90°
95°
100°
20°
15°
10°
5°
0°

to form an extensive blanket and was called *red clay*, although actually most of it is some shade of brown. Now it is generally called *brown clay.* In the shoaler areas, they found what they called *ooze*, consisting of skeletal remains of small organisms, such as foraminifers, radiolarians, and diatoms. Off large rivers, they discovered muds of various colors. Other types of sediment are mentioned briefly in the *Challenger* reports.

Fig. 11-9 (page 196). Bengal Fan channels extending south of the Ganges Delta into depths of 5,000 meters. From J. Curray and D. G. Moore, *Geological Society of America Bulletin*, 1971.

The *Challenger* Expedition discoveries remained standard doctrine until, in the 1940s, oceanographers began to obtain longer cores. Charles Piggot of the Carnegie Institute invented a gun that shot a core barrel into the bottom for about 10 feet (3 m). Cores taken by this method allowed Bramlette and Bradley (1940) to find from a crossing of the North Atlantic that the deep-sea sediments were often stratified with sand layers alternating with the usual deep-sea oozes. The Piggot coring method was abandoned after the device exploded at the surface and almost put a hole through the hull of the Woods Holes's ship *Atlantis*. As a result of one lowering where the explosive did not go off just as good a core was obtained merely from the heavy weight. Next was the so-called piston corer invented by Börje Kullenberg of Sweden. The piston is released near the bottom and the core falls free past the piston, penetrating as much as 100 feet (34 m) into the bottom in exceptional circumstances (Zenkevitch, 1955). This device was used in the world-encircling Swedish *Albatross* Expedition under the leadership of Hans Pettersson (1953). The numerous long cores were described by Gustaf Arrhenius (1952), now of Scripps Institution. His results confirmed those of the Piggot cores, showing that sediments of the deep ocean may change radically along their length.

One of the results of the *Albatross* Expedition was the appreciation of the changes between layers with relatively warm and relatively cool-water foraminifera, which are often found along the length of cores. This was further indicated by many Woods Hole expeditions in which Fred Phleger, now of Scripps Institution, participated along with David Ericson, now of Lamont-Doherty Geological Observatory. Many others continued the work.

The study of the siliceous radiolarians from deep-sea cores has been particularly interesting. William Riedel of Scripps Institution (1971) has found that radiolarian oozes in the Pacific often contain fossils from sediments near or just below the sea floor that are millions of years old, going back as far as the Eocene. It is now possible to determine the water temperatures at which the foraminifera were living near the surface. This method, called Oxygen 16-18, was developed by Harold Urey (1947), and has been used extensively by Cesare Emiliani (1966), of the School of Marine and Atmospheric Sciences, Miami University. It has shown the world-wide importance of ocean temperature changes during glacial and interglacial stages.

Deep-sea sands. The Piggot cores obtained in high latitudes of the Atlantic contained many sand layers, which were generally attributed to glacial stages when icebergs were carrying sands and gravels far from the ice fronts. Somewhat later, cores from other parts of the oceans indicated that deep-sea sands were by no means confined to the borders of the great ice caps. In fact, sand layers proved to be common in cores from the continental rise and even from the deep basins beyond. However, the layers were almost entirely missing where a submarine ridge had intervened between the core locality and the lands. As the importance of turbidity currents became more and more confirmed, the explanation for these sand layers was made evident. Given even the gentlest of marine slopes, the turbidity currents keep on moving to the greatest depths, still with sufficient velocity to carry coarse sediments. Scientists have found sand layers derived from Brazil in mid-Atlantic (Damuth and Kumar, 1975).

Manganese Nodules, The Eldorado of the Future from the Deep Seas

When the *Challenger* Expedition brought up the first ferromanganese nodules from the deep sea floor a century ago, no one surely imagined that these were anything but a curiosity. It was not until Mero (1965) began to enlighten us that mineral economists began to think seriously of the possibility that the manganese nodules (with their high content of nickel, copper, and cobalt) might become a major world source of these metals. Bottom photography was perhaps one of the first strong arguments that we might begin to scrape the ocean bottom with dredges and bucket-chains on such a large scale as to make these nodules an economic resource. In many places the photographs showed that the sea floor was almost covered with small rounded nodules (Fig. 11-10). Now, many companies are in the process of developing equipment which may soon bring this dream to reality.

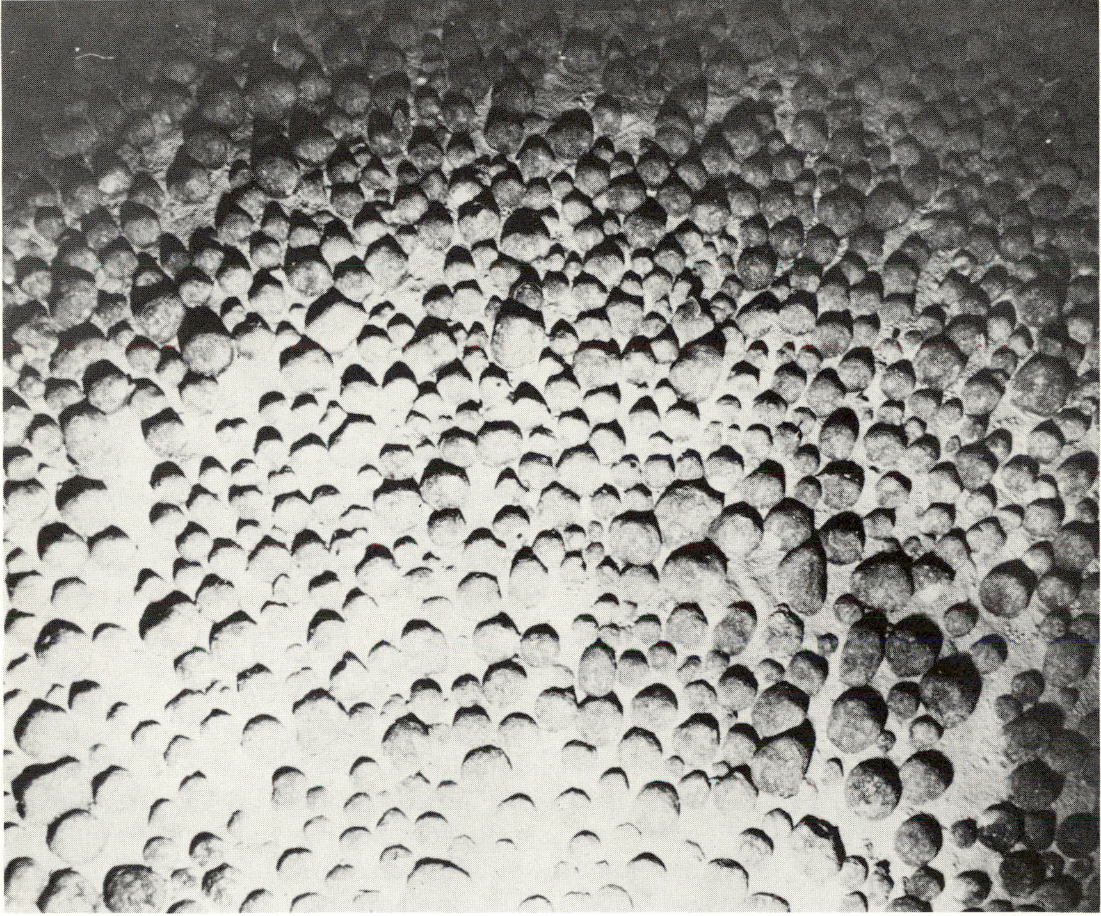

Fig. 11-10. Bottom photograph at 5000 meters showing the abundance of manganese nodules. Location, south equatorial Pacific. From Lamont-Doherty Geological Observatory.

Beginning in 1972, under the coordination of Robert Gerard of Lamont-Doherty Geological Observatory, an inter-university program was started that is attempting to coordinate the various studies of manganese nodules. A preliminary report by the National Science Foundation (1973), the sponsor of the project, gives information on the nature of these curious oval nodules that line so much of the deep ocean floor. This also explains some of the difficulties that will have to be overcome before large mining of the deep sea floor is undertaken. The extensive work by Horn and others (1973), also of Lamont-Doherty, gives much new information concerning the richness of phosphorite nodules in nickel and copper in different parts of the oceans. It is interesting to consider that large scale mining of the manganes nodules would give us as byproducts about 300% of our present annual consumption of manganese, about 200% of that of cobalt, and many times that of titanium, vanadium, and other rare metals. However, there is much investigation to be accomplished before one can expect to see a general change in mining operations from the lands to the deep sea.

Under the Ocean Floor

By estimating the rate at which the ocean oozes and the brown clays are being deposited, by making reasonable corrections for earlier ages, and by assuming about 2 billion years for the ocean age, Phillip Kuenen (1946) estimated that there should be an average thickness of ocean sediment of 10,000 feet (3048 m). This, of course, was before plate tectonics was well established and before deep ocean drilling. Now we have a good idea of the actual thickness as the result of the many drillings and profiles made by sound reflection and refraction. The rough estimates are 2000 feet (610 m) for the Pacific and 3300 feet (1000 m) for the Atlantic. These figures are reasonable when one takes into account the opening of the new oceans and of the subduction of much, perhaps all, of the Pacific crust in the treadmill process of sea-floor spreading.

Maps have now been prepared to show the age of the oldest sediments under various parts of the ocean basins (Fig. 11-11). It is interesting to find that Jurassic (about 170 million years old) is the oldest found under any of the deep-ocean floors. Also, in general, the sediments overlying the lava crust show an increasing age going away from the Mid-Atlantic Ridge and from most of the rest of the mid-ocean ridges. Some complications are found in the western North Pacific, where additional spreading zones seem to exist.

The Deep Sea Drilling Project has provided us with a wealth of new information about the sub-ocean. It is a joint project of nine United States marine laboratories and is supported by six other nations, with the entire project administered by Scripps Institution of Oceanography.

Among the interesting discoveries of the drilling project have been the salt domes of the Gulf of Mexico and the salt layers under the Mediterranean and various parts of the marginal Atlantic; the gradation downward in many drillings from relatively deep deposits, like the brown

Fig. 11-11. Age of ocean crust in the Pacific. Information from Deep-Sea Drilling and dredging operations of Scripps Institution of Oceanography. From E. L. Winterer, *American Association of Petroleum Geologists Bulletin,* 1975.

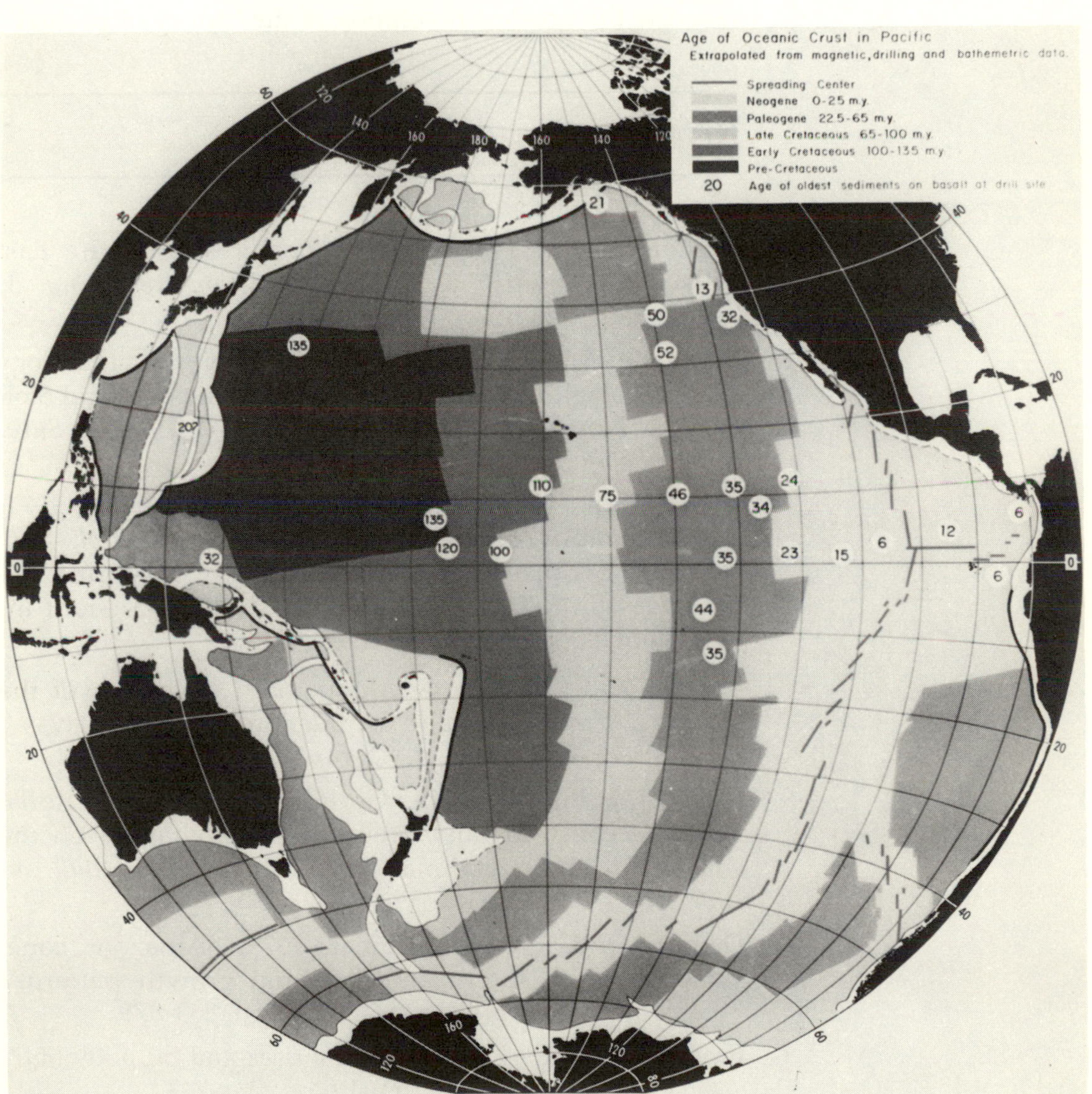

clays, to relatively shallow deposits, like the foraminiferal oozes; the discovery of thick chert layers, mostly of organic origin, that made drilling difficult; and the frequency of penetrating sand and even gravel layers, suggestive of turbidity currents or in some cases of large deepening of the ocean floors. A major ridge in the Indian Ocean now more than a half-mile (0.8 km) deep has yielded coal and layers of shallow-water shells, suggesting ancient islands with swamps and lagoons. Most of these discoveries have tended to strengthen the exciting crustal migration hypothesis. Thus, we conclude with still more evidence favoring the greatest revolution in the history of geology.

References

Arrhenius, G., 1952. "Sediment cores from the East Pacific." *Swedish Deep Sea Expedition Rept., no. 5,* Göteborg, 186 pp.

Bramlette, M. N., and W. H. Bradley, 1940. "Lithology and geologic interpretations. Pt. 1. In Geology and Biology of North Atlantic Deep-Sea Cores." *U.S. Geol. Surv. Prof. Paper 196,* pp. 1-24.

Chase, T. E., H. W. Menard, and J. Mammerickx, 1971. "Topography of the North Pacific map." *Geol. Data Center,* Scripps Inst. Oceanog. and Inst. Marine Resources. IMR Tech. Rept. Series, TR-17; Sea Grant Institutional Program GH-112.

Curray, J. R., and D. G. Moore, 1971. "Growth of the Bengal deep-sea fan and denudation in the Himalayas." *Geol. Soc. Amer. Bull.,* v. 82, no. 3, pp. 563-572.

Damuth, J. E., and M. A. Gorini, 1976. "The Equatorial Mid-Ocean Canyon: A relic deep-sea channel on the Brazilian continental margin." *Geol. Soc. Amer. Bull.,* v. 87, no. 3, pp. 340-346.

Damuth, J. E., and N. Kumar, 1975. "Amazone cone: morphology, sediments, age, and growth pattern." *Geol. Soc. Amer. Bull.,* v. 86, no. 6, pp. 863-878.

Emery, K. O., 1969. "Continental rises and oil potential." *Oil and Gas Jour.*, v. 67, no. 19, pp. 231-243.

Emiliani, C., 1966. "Isotopic paleotemperatures." *Science,* v. 154, no. 3751, pp. 851-857.

Griggs, G. B., A. G. Carey, Jr., and L. D. Kulm, 1969. "Deep-sea sedimentation and sediment-fauna interaction in Cascadia Channel on Cascadia Abyssal Plain." *Deep-Sea Res.*, v. 16, no. 2, pp. 157-170.

Hamilton, E. L., 1956. "Sunken islands of the Mid-Pacific Mountains." *Geol. Soc. Amer. Mem. no. 64,* 97 pp.

Heezen, B. C., 1956. "The origin of submarine canyons." *Sci. Amer.*, v. 195, no. 2, pp. 36-41.

Heezen, B. C., M. Tharp, and M. Ewing, 1959. "The Floors of the Oceans. 1. The North Atlantic. Text to Accompany the Physiographic Diagram of the North Atlantic." *Geol. Soc. Amer.,* spec. pap. 65, 122 pp.

Hess, H. H. 1946. "Drowned ancient islands of the Pacific basin." *Amer. Jour. Sci.*, v. 244, pp. 772-791.

Hollister, C. D., R. D. Flood, D. A. Johnson, P. Lonsdale, and J. B. Southard, 1974. "Abyssal furrows and hyperbolic traces on the Bahama Outer Ridge." *Geology*, v. 2, no. 8, pp. 395-400.

Horn, D. R., B. M. Horn, and M. N. Delach, 1973. "Ocean manganese nodules, metal values, and mining sites." *Tech. Rept. no. NSF GX 33616,* Intl. Decade of Ocean Exploration. Nat. Sci. Found., Washington, D.C., 57 pp.

Hurley, R. J. 1960. *The Geomorphology of Abyssal Plains in the Northeast Pacific Ocean.* Ph.D. thesis, Scripps Inst. Oceanog., Univ. Calif., San Diego, 173 pp.

Kuenen, P. H., 1946. "Rate and mass of deep sea sedimentation." *Amer. Jour. Sci.*, v. 244, pp. 563-572.

Larson, R. L., 1970. *Near-botton Studies of the East Pacific Rise Crest and Tectonics of the Mouth of the Gulf of California.* Unpublished Ph.D. Thesis. Scripps Inst. Oceanog., Univ. Calif., San Diego, 164 pp.

Matthews, J. L., B. C. Heezen, R. Catalano, A. Coogan, M. Tharp, J. Natland, and M. Rawson, 1974. "Cretaceous drowning of reefs on mid-Pacific and Japanese guyots." *Science,* v. 184, no. 4135, pp. 462-464.

Menard, H. W., 1955. "Deformation of the northeastern Pacific Basin and the west coast of North America." *Geol. Soc. Amer. Bull.*, v. 66, pp. 1149-1198.

Menard, H. W., and R. S. Dietz, 1952. "Mendocino Submarine Escarpment." *Jour. Geol.*, v. 60, no. 3, pp. 266-278.

Menard, H. W., and S. M. Smith, 1966. "Hypsometry of ocean basin provinces." *Jour. Geophys. Res.*, v. 71, no. 18, pp. 4305-4325.

Mero, J. L., 1965. *The Mineral Resources of the Sea.* Elsevier Oceanography Series, Elsevier Publ. Co., N.Y., 312 pp.

Murray, Sir John, and A. F. Renard, 1891. "Deep-Sea Deposits Based on the Specimens Collected during the Voyage of H. M. S. *Challenger* in the Years 1872-1876, *"Challenger" Reports.* Longmans, London, 525 pp. (reprint; Johnson, London, 1965).

National Science Foundation, Wash., D.C. (International Decade of Ocean Exploration, Seabed Assessment Program), 1973. Inter-University Program of Research on Ferromanganese deposits of the Ocean Floor. Phase I Report.

Riedel, W. R., 1971. "The occurrence of the Pre-Quaternary Radiolaria in deep-sea sediments." In *Micropaleontology of Oceans.* B. M. Funnell and W. R. Riedel, eds., Cambridge Univ. Press, pp. 567-594.

Scholl, D. W., M. N. Christensen, R. von Huene, and M. S. Marlow, 1970. "Peru-Chile Trench sediments and sea-floor spreading." *Geol. Soc. Amer. Bull.*, v. 81, no. 5, pp. 1339-1360.

Urey, H. C. 1947. "The thermodynamic properties of isotopic substances." *Jour. Amer. Chem. Soc.*, pp. 562-581.

Vacquier, V., 1972. *Geomagnetism in Marine Geology.* Elsevier Oceanography Series, 6. Elsevier Publ. Col, N.Y., 185 pp.

Vine, F. J., 1968. "Magnetic anomalies associated with mid-ocean ridges." In *The History of the Earth's Crust.* R. A. Phinney, ed., Princeton Univ. Press, N.J., pp. 73-89.

Winterer, E. L., 1973. "Sedimentary facies and plate tectonics of equatorial Pacific." *Amer. Assoc. Petrol. Geol. Bull.*, v. 57, no. 2, pp. 265-282.

Zenkevitch, L. A., 1955. "Importance of deep-sea research." *Trudi Inst. Okeanol.*, 12, pp. 5-15. Trans. by Admiralty Center for Scientific Inform. and Liaison.

Suggested Supplementary Reading

Heezen, B. C., and C. Hollister, 1971. *The Face of the Deep.* Oxford Univ. Press, N.Y. and London, 659 pp.

Heezen, B. C., and I. D. MacGregor, 1973. "The evolution of the Pacific." *Sci. Amer.*, v. 229, no. 5, pp. 102-112.

Heitzler, J. R., and W. B. Bryan, 1975. "The floor of the Mid-Atlantic rift." *Sci. Amer.*, v. 233, no. 2, pp. 78-90.

MacIntyre, F., 1970. "Why the sea is salt." *Sci. Amer.*, v. 223, no. 5, pp. 104-115.

Mero, J. L., 1960. "Minerals on the ocean floor." *Sci. Amer.*, v. 203, no. 6, pp. 64-72.

Rona, P. A., 1973. "Plate tectonics and mineral resources." *Sci. Amer.*, v. 229, no. 1, pp. 86-95.

Index